Crystal Structures

Lattices and Solids in Stereoview

T0227738

SICILLLUM MAIORAT CIVITATIS CICESTRIE
These two Mayoral Seals of the Roman city of Chichester
in 1502 and 1530 AD were design *motifs* for our colophon

"*Happiness in a thinking man is to have fathomed what can be fathomed,
and quietly to reverence the unfathomable.*"

Johann Wolfgang von Goethe (1749-1842)

"*Never is there either work without reward, or reward without work
being expended.*"

Livy (*circa* 20 BC)

MARK LADD, DSc(*Lond.*)

Mark Ladd hails from Porlock in Somerset and subsequently his family moved to Bridgwater, where his initial education was at Dr John Morgan's school there. Before going up to university he worked for three years in the Analytical Chemistry Laboratory of the Bridgwater Royal Ordnance Factory, followed by three years' National Service with the Royal Army Ordnance Corps. On demobilization he studied chemistry at London University and obtained Bsc (Special) degree in 1952. He then worked until 1955 in the General Electric Company in Wembley in the Ceramic and Refractories Division of their research laboratories, during which service he obtained his MSc degree in crystallography from London University. In 1955, he became lecturer at Battersea Polytechnic later renamed Battersea College of Advanced Technology and later becoming the University of Surrey. He was awarded his PhD by London University for research on crystallogaphy of the triterpenoids with particular reference to the crystal and molecular structure of euphadienol. In 1979 he was admitted to the degree of DSc by the London University for research contributions in crystallography and solid-state chemistry. He was until recently Reader in the Department of Chemistry at Surrey University.

Mark Ladd is the author or co-author of the many books: *Analytical Chemistry: Radiochemistry: Introduction to Physical Chemistry: Direct Methods in Crystallography: Structure Determination by X-Ray Crystallography 3rd edition: Structure and Bonding in Solid-State Chemistry: Symmetry in Molecules and Crystals*, and *Chemical Bonding in Solids and Fluids*, (the last three with Ellis Horwood Limited). His other recent books were *Introduction to Physical Chemistry 3rd edition* (Cambridge University Press) and *Symmetry and Group Theory in Chemistry* (Horwood Publishing Limited). He has published more than one hundred research papers on crystallography and in the energetics and solubility of ionic compounds.

His other activities include music: he plays the viola and the double bass in orchestral and chamber ensembles, and has performed the solo double bass in the Mozart's *Serenata Notturna* and the Saint-Saens's *Carnival of Animals*. He has been exhibitor, breeder and judge of Dobermanns, and has trained Dobermanns in obedience and he has written the successful *Dobermanns: An Owner's Companion*, (Crowood Press and Howell Book House, New York). Currently he is working in conjunction with the Torch Trust in the computer transcriptions of African language Bibles into braille, and has completed the whole of the Chichewa (Malawi) and Bemba (Zambia) Bibles.

Mark Ladd is married with two sons: one is Professor in the Department of Chemical Engineering at the University of Florida in Gainsville, and the other is the vicar of St Luke's Anglican Church in Brickett Wood, St Albans, Hertfordshire. He lives in Farnham, Surrey with his wife and one Dobermann.

Crystal Structures
Lattices and Solids in Stereoview

Mark Ladd, DSc(Lond), FRSC, FInstP
Department of Chemistry
University of Surrey
Guildford

Horwood Publishing
Chichester

First published in 1999 by
Reprinted 2002

HORWOOD PUBLISHING LIMITED
International Publishers
Coll House, Westergate, Chichester, West Sussex, PO20 6QL
England

COPYRIGHT NOTICE
All Rights Reserved. No part of this publication may be reproduced, stored in a retrieval system, or transmitted, in any form or by any means, electronic, mechanical, photocopying, recording, or otherwise, without the permission of Horwood Publishing, International Publishers, Coll House, Westergate, Chichester, West Sussex, England

© M. F. C. Ladd, 1999

British Library Cataloguing in Publication Data
A catalogue record of this book is available from the British Library

ISBN 1-898563-63-2

Table of contents

Table of contents

Table of contents

Table of contents

Preface

This book presents and discusses those common crystal structures that would be encountered by students reading chemistry, or any other subject within which chemistry forms a significant component. The work is applicable to undergraduate study modules: the year is somewhat dependent upon the structure of any particular course, but corresponds most generally to a second-year level. Thus, some groundwork in topics such as elementary thermodynamics and introductory atomic theory is presumed, together with the ability to handle first and second differentials of a single variable and to distinguish between a scalar and a vector.

Structures have been divided, for convenience, into four groups, determined by the type of interparticle bonding force responsible for cohesion of the structural units in the solid state. In addition to the descriptions of the structures, the theory of bonding is presented for each of the groups of solids.

The crystal structures, and lattices, are illustrated by copious stereoviews, and instructions for their correct viewing and for constructing a stereoviewer are provided. In the text, the structures are discussed with respect to crystallographic data, including space groups, molecular geometry and molecular symmetry. A comprehensive coverage of space groups was beyond the scope of the present book. However, a certain number of common space groups has been included, and the reader will find it helpful to refer to the International Tables for Crystallography Volume A (Volume I in its previous edition) as the definitive work on this topic.

An introduction to symmetry is provided in terms of point groups, lattices and space groups, insofar as is required for an understanding of the structures themselves. The Hermann-Mauguin (international) notation is used throughout the book. The point groups $4/mmm$ and $6/mmm$ are frequently written in this manner. In this book they (and the space groups belonging to them) have been written as $4/m\,mm$ and $6/m\,mm$, so as to emphasize that $4/m$ and $6/m$ each correspond to *one* direction in their respective point-group symbols, and mm to the two other directions in the symbols.

The mathematics required in studying the subject matter of this book is commensurate with that encountered in an A-level mathematics course by the modern chemistry student. Other relevant mathematical topics are provided in appendices.

Each chapter is provided with worked examples (with answers), and more detailed problems for which tutorial solutions are provided. The solving of certain problems are facilitated by computer programs, and a small set of programs can be downloaded as *.EXE* files under the internet address **www.horwood.net/publish.** The final chapter has been devoted to the technique of solving problems, and to the use of programs on the internet. The reader may wish to peruse this chapter at an earlier stage than is suggested by its point of inclusion.

The author is pleased to express his thanks to Dr John Burgess of the Chemistry Department of the University of Leicester for helpful suggestions in the early stages of preparation, to Dr D C Povey of the Chemistry Department of the University of Surrey for reading the manuscript in proof, to those publishers and authors for permission to reproduce those illustration that carry appropriate acknowledgements, and to Ellis Horwood MBE FRSC of Horwood Publications Limited for encouragement and for bringing the book to a state of completion.

Farnham, 1999   Mark Ladd

Physical constants and other numerical data

These data have been selected, or derived, from the compilation of E R Cohen and B N Taylor, *J Phys Chem Ref Data* **17**, 1795-1803 (1988). The values are reported in SI units, and some conversions to other units are given. The reader should refer to the original paper for these and other data at their full known precisions.

Speed of light in a vacuum (defined)	c	2.9979×10^8	$m\ s^{-1}$
Permittivity of a vacuum (defined)	ε_0	8.8542×10^{-12}	$F\ m^{-1}\ (kg^{-1}\ m^{-3}\ s^4\ A^2)$
Elementary charge	e	1.6022×10^{-19}	$C\ (A\ s)$
Planck constant	h	6.6261×10^{-34}	$J\ Hz^{-1}\ (kg\ m^2\ s^{-3})$
Rest mass of electron	m_e	9.1094×10^{-31}	kg
Avogadro constant	L	6.0221×10^{23}	mol^{-1}
Atomic mass unit	u	1.6605×10^{-27}	kg
Boltzmann constant	k_B	1.3807×10^{-23}	$J\ K^{-1}\ (kg\ m^2\ s^{-2}\ K^{-1})$
Molar gas constant	R	8.3145	$J\ K^{-1}\ mol^{-1}$
		8.2058×10^{-2}	$dm^3\ atm\ K^{-1}\ mol^{-1}$ (see below)
Faraday	F	9.6485×10^4	$C\ mol^{-1}\ (A\ s\ mol^{-1})$
Ice-point temperature	T_{ice}	273.15	K

1F (Farad) $= 1\ C\ V^{-1}$.
1Å (ångström) $= 10$ nm.
1 D (Debye) $= 3.3356 \times 10^{-30}\ C\ m$.
1 eV (electronvolt) $= 1.6022 \times 10^{-19}\ J$ $\equiv 96.485$ kJ mol^{-1}.
1 atm (atmosphere) $= 101325$ Pa (N m^{-2}, or kg m^{-1} s^{-2}) $= 760$ Torr $= 760$ mmHg.

Prefixes in common use

femto	pico	nano	micro	milli	centi	deci	kilo	mega	giga
f	p	n	μ	m	c	d	k	M	G
10^{-15}	10^{-12}	10^{-9}	10^{-6}	10^{-3}	10^{-2}	10^{-1}	10^3	10^6	10^9

List of symbols

A	A-face centred unit; Madelung constant; area
a	Unit cell vector
a	Unit cell dimension; a-glide plane; activity
a_0	Bohr radius for hydrogen
B	B-face centred unit cell
C	C-face centred unit cell; molar heat capacity
b	Unit cell vector
b	Unit cell dimension; b-glide plane
c	Unit cell vector
c	Speed of light in a vacuum; unit cell dimension; centred unit cell (two dimensions); c-glide plane; LCAO constant
cr	Crystal
D	Debye
D	Diffusion coefficient
D_e	Theoretical dissociation energy (D_0 – zero-point energy)
D_0	Experimental dissociation energy; limiting diffusion coefficient
d	Differential operator
d	Bond distance; d-orbital; sub-shell descriptor
∂	Partial differential operator
∇^2	Laplacian operator ($\partial^2/\partial x^2 + \partial^2/\partial y^2 + \partial^2/\partial z^2$)
E	Energy
E_F	Fermi energy
e	Electronic charge
exp	Exponential function
F	All-face centred unit cell
f	Sub-shell descriptor; Fermi-Dirac distribution function; general function
G	Gibbs' free energy
g	Gas
g	Even (parity); density of states function; Goldschmidt factor
$\mathbf{H}$	Hamiltonian operator
H	Electronic Hamiltonian operator
H	Enthalpy; Coulomb integral
h	Planck constant; hybrid orbital
$\hbar$	('Cross-h') $h/2\pi$
I	Body-centred unit cell
i	Unit vector
i	$\sqrt{-1}$
J	Joule
K	Equilibrium constant; shell descriptor
k	Wave vector (in reciprocal space)
$k, |k|$	Magnitude of wave vector k
k_B	Boltzmann constant
k_F	Radius of Fermi sphere
L	Avogadro constant; shell descriptor
l	Angular momentum quantum number
ln	Naperian logarithm ($\log_e$)

m	Reflection (mirror) symmetry plane; mass; molality
m_e	Rest mass of electron
m_l	Angular momentum (magnetic) quantum nummber
m_p	Mass of proton
m_s	Spin quantum number resolved along z axis ($\pm\frac{1}{2}$)
M	General symbol for a metal; shell descriptor; molar mass (relative)
N	Second general symbol for a metal; normalization constant; shell descriptor
n	Number of moles; n-glide plane; principal quantum number; coordinate in reciprocal space
o	Odd (parity)
P	Primitive unit cell (three dimensions)
p	Primitive unit cell (two dimensions); dipole moment; pressure; polarizing power; p-orbital; sub-shell descriptor; band descriptor
q	Ionic charge
R	Rotation symmetry axis; primitive rhombohedral unit cell; radius ratio; radial wavefunction
$\bar{R}$	Roto-inversion symmetry axis
$\boldsymbol{R}$	Universal gas constant
$\boldsymbol{r}$	Position vector
r	Interatomic distance; radius of a species; spherical polar coordinate; correlation coefficient
S	Entropy; overlap integral
s	s-Orbital; sub-shell decsriptor; band descriptor; spin quantum number
T	Temperature
t	Translation repeat in lattice
U	Intrinsic energy; lattice energy
u	Reference axis (hexagonal system); electrostatic energy per pair of ions
V	Volume (of unit cell); poyential energy
v	Speed
X	General symbol for a non-metal
x	Reference axis; fractional coordinate in unit cell; general distance
y	Reference axis; fractional coordinate in unit cell
Z	Atomic number
z	Reference axis; fractional coordinate in unit cell
α	Coulomb integral; polymorph indicator; alloy phase descriptor
α	Interaxial angle; polarizability; thermal expansivity
α'	Volume polarizability
β	Resonance integral; polymorph indicator; alloy phase descriptor
β	Interaxial angle
γ	Alloy phase descriptor
γ	Interaxial angle; activity coefficient
χ	Electronegativity
δ	Infinitesimal amount; partial charge; Lennard-Jones collision diameter
δ_{ij}	Kronecker delta

Δ	Difference between two quantities
ϵ	Alloy phase descriptor
ε	Electrostatic bond strength; relative permittivity
ε_0	Permittivity of a vacuum
ε_{LJ}	Lennard-Jones energy parameter
η	Alloy phase descriptor
ε_0	Permittivity of a vacuum
Θ	Debye temperature
θ	Spherical polar coordinate
κ	Magnetic susceptibility; isothermal compressibility
λ	Fractional ionic character
μ	Reduced mass
ν	Frequency
π	Ratio of the circumference of a circle to its diameter (3.141592···); bonding π-orbital
π^*	Antibonding π-orbital
ρ	Electron density; electrical resistivity; exponent in repulsion potential energy function
σ	Bonding σ-orbital
σ	Screening constant; standard deviation
σ^*	Antibonding σ-orbital
τ	Infintesimal volume
ϕ	Molecular wavefunction; spherical polar coordinate; general angle
ψ	Atomic wavefunction
ψ^*	Atomic wavefunction conjugate to ψ

Superscripts

$\ominus$	Standard state indicator
$-$	Average value
$+$	Positive nature of main symbol
$-$	Negative nature of main symbol

Subscripts

d	Dissolution; dipole; diffusion
E	Electrostatic
e	Equilibrium; electron affination
f	Formation
h	Hydration
i	Ionization; ionic
id	Induced dipole
ig	Ion-gas
l	Lattice
m	Molar; melting-point
p	Pressure
r	Relative
s	Sublimation; solvation
v	Volume
$+$	Positive nature of main symbol
$-$	Negative nature of main symbol

1

Lattices and their properties

1.1 INTRODUCTION

A short paper published in 1926[1] discussed the geometry of stereoscopic views (stereoviews), and gave details for their construction. Subsequently, stereoviews began to be used to illustrate three-dimensional structures. Nowadays, this technique is quite commonplace, and computer programs exist that prepare the two views needed for producing a three-dimensional image of a crystal or molecular structure.

This book discusses the geometrical bases of crystal structures, the Bravais lattices, and a range of common structures. We shall consider also the theory of the bonding in these structures; stereoviews are employed to illustrate both the structures and the Bravais lattices. We begin with a study of lattices, but first a word on viewing stereoscopic illustrations, and a brief survey of the symmetry of molecules and crystals.

1.2 USING STEREOVIEWS

The two diagrams of a given object, necessary to form a three-dimensional visual image, must correspond to the views seen by the eyes in normal vision. Correct viewing of a stereoscopic diagram requires that each eye sees only the appropriate half of the complete illustration, and there are two ways in which it may be accomplished.

The simplest procedure is with a **stereoviewer**. A supplier of a stereoviewer that is relatively inexpensive is:

Taylor-Merchant Corporation, 212 West 35th Street, New York, NY 10001, USA
(Tel: 001 212 757 7700; Fax: 001 212 695 1265).

The pair of drawings is viewed directly with the stereoviewer, whereupon the three-dimensional image appears centrally between the two given diagrams.

An alternative procedure involves training the unaided eyes to defocus, so that each eye sees only the appropriate diagram. The eyes must be relaxed and look straight ahead. This process may be aided by holding a white card edgeways between the two drawings. It may be helpful to close the eyes for a moment, then to open them wide and allow them to relax without consciously focussing on the diagram.

Finally, we give instructions whereby a simple stereoviewer can be constructed with ease. A pair of planoconvex or biconvex lenses each of the same focal length, approximately 100 mm, and diameter approximately 30 mm, is mounted between two opaque cards such that the centres of the lenses are approximately 63 mm apart. The card frame must be so shaped that the lenses may be brought close to the eyes. Figure 1.1 illustrates the construction of the stereoviewer.

1.3 SYMMETRY OF MOLECULES AND CRYSTALS

In studying lattices and structures we shall be confronted by considerations of **symmetry**, first with finite bodies, molecules and crystals, and then in extended arrays of chemical units, crystal structures. We shall consider these topics briefly, and refer the reader to other sources that discuss symmetry more extensively[2-6].

We define a symmetry **operation** as *an action that moves a body into a position that is indistinguishable from its position before that operation*. The symmetry operation

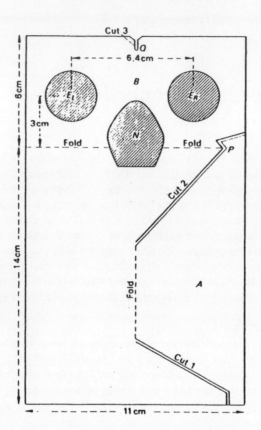

Fig. 1.1. Simple stereoviewer. Cut out two pieces of card as shown, and discard the shaded portions. Make cuts along the double lines. Glue the two cards together with lenses E_L and E_R in position, fold the portions A and B backwards, and fix the projection P into the cut at Q. Strengthen the fold with a strip of 'Selotape'. View from the side marked B. (This type of viewer is marketed by the Taylor-Merchant Corporation, New York.). It may help to obscure a segment on each lens of depth *ca* 30% of the lens diameter, closest to the nose region.

reveals the symmetry present in the body, and may be considered to take place with respect to a symmetry **element**. A symmetry element is a geometrical entity, a point, a line or a plane, which is associated with its corresponding symmetry operation. Symmetry elements are, strictly, conceptual but, as long as we remember this fact, it is convenient to accord them a sense of reality: none of the symmetry elements of a finite body involves the operation of translation.

1.3.1 Symmetry operations and symmetry elements
For our present purposes, we need to discuss three symmetry operations, or symmetry elements, those of *rotation R*, *reflection m*, and *roto-inversion* $\overline{R}$ (which includes the centre of symmetry).

Rotation
A body exhibits an R-fold rotational symmetry axis if an operation of $(360/R)°$ about that axis brings the body into an orientation that is indistinguishable from that before the operation. In principle R may take any value from 1 to infinity; in crystals, the permitted

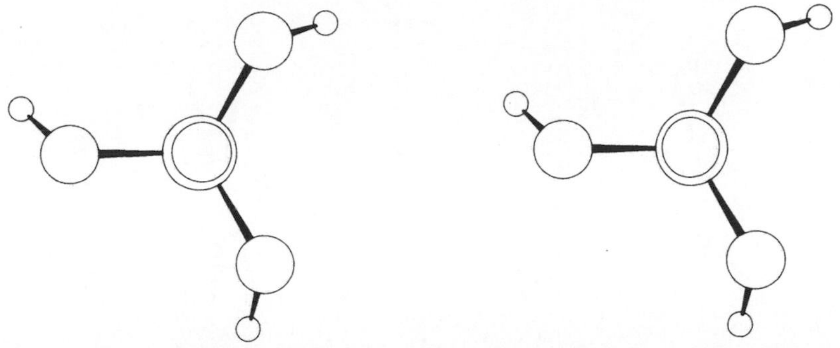

Fig. 1.2 Stereoview of the molecule of orthophosphoric acid H_3PO_4; the circles in order of decreasing size represent P, O and H species. The three-fold axis is along the P=O bond.

values of R are 1, 2, 3, 4 and 6 only, as we shall show in the discussion on lattices. All bodies possess trivial, one-fold rotation, also called **identity** symmetry. Rotation is the one operation that we can perform physically on an object; the others must be imagined. Figure 1.2 is a stereoview of the molecule of orthophosphoric acid, which shows rotational symmetry along the P=O bond.

Reflection
A body exhibits reflection symmetry m if it can be imagined to be divided into halves that are related to each other as an object is to its mirror image; the plane of division is a reflection symmetry (**mirror**) plane. Reflection across this plane would leave the body indistinguishable from its appearance before the operation. Figure 1.3 is a stereoview of the chlorobenzene molecule, C_6H_5Cl: it exhibits m-planes both as the molecular plane and as that passing through the chlorine and opposite carbon atoms; the two m-planes intersect in a two-fold rotation axis. Mirror symmetry will be familiar to those who have studied the optical activity of chemical species.

Roto-inversion
A body exhibits a roto-inversion axis $\bar{R}$, also called an inversion axis, if the combined actions of rotation through $(360/R)°$ followed by inversion through a point on the $\bar{R}$ axis leaves the body indistinguishable from its original position; the two motions constitute a

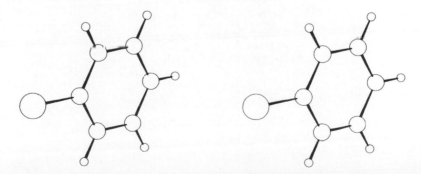

Fig. 1.3 Stereoview of the molecule of chlorobenzene C_6H_5Cl; the circles in order of decreasing size represent Cl, C and H species. The m-planes are the molecular plane and that passing through the Cl and C(4) atoms.

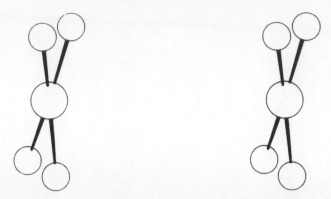

Fig. 1.4 Stereoview of the molecule of thorium tetrabromide $ThBr_4$; the circles in order of decreasing size represent Br and Th species. The $\bar{4}$ axis is the vertical axis in the diagram.

single symmetry operation. Again, there is, in principle, no limit on the value that R can take in a roto-inversion axis. When R is an odd integer, however, the point of inversion corresponds to a centre of symmetry $\bar{1}$. Figure 1.4 shows a stereoview of thorium tetra-bromide $ThBr_4$; the vertical axis is $\bar{4}$. The tetrahedron, which we construct in Example 1.1(b) below, is another body that exhibits four-fold roto-inversion axes; there are three of them. As in the case of thorium tetrabromide, there is more symmetry than that of roto-inversion alone in the tetrahedron.

1.3.2 Reference axes

We use a system of reference axes, x, y and z, which are chosen parallel to important symmetry directions in a given body and form a right-handed set. Figure 1.5 illustrates these axes drawn in a crystal; the sequence $x \rightarrow y \rightarrow z$ simulates a right-handed screw motion. Where the symmetry of the body requires that these axes are not mutually perpendicular, the interaxial angles are denoted α between y and z, β between z and x, and γ between x and y.

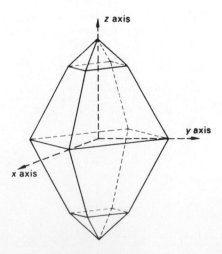

Fig. 1.5 Idealized (tetragonal) crystal, showing the crystallographic x, y and z axes.

EXAMPLE 1.1. We make models of a cube and a tetrahedron, and investigate their symmetry. (a) Cube. On a thin card, draw a square of side, say, 50 mm. On each side of this square draw another identical square. Lightly score the edges of the first square, fold the other four squares in the same sense to form the faces of a cube, and fasten with Selotape. There is an advantage in leaving the sixth face of the cube open, as we shall see, but we shall imagine its presence as needed. (b) Tetrahedron. On a similar card, draw an equilateral triangle of side $a\sqrt{2}$, where a is the length chosen for the side of the cube in (a). On each side of the triangle, draw another identical triangle. Lightly score the edges of the first triangle, fold the other three triangles in the same sense to meet at an apex, and fasten with 'Selotape'.

Symmetry. Imagine that the x, y and z reference axes lie normal to three non-coplanar faces of the cube. Determine the symmetry elements in relation to these axes. It may help to number the corners 1–8, but remember that the numbers are not part of the cube. Repeat the exercise with the tetrahedron. It will be found that the tetrahedron may be placed inside the cube with its edges forming the six face diagonals of the cube. The reference axes are now in position, and the symmetry elements common to both polyhedra coincide.

1.3.3 Point groups in general

Table 1.1 Symbols of the crystallographic point groups

General symbol	Triclinic	Monoclinic (z unique)	Trigonal	Tetragonal	Hexagonal	Cubic
R	1	2	3	4	6	23
$\overline{R}$ (even)	-	$m (= \overline{2})$	-	$\overline{4}$	$\overline{6}$	-
R+centre	$\overline{1}$	$2/m$	$\overline{3}$	$4/m$	$6/m$	$m3$

	Monoclinic (y unique)	Orthorhombic				
$R2$	2	222	32	422	622	432
Rm	m	$mm2$	$3m$	$4mm$	$6mm$	-
$\overline{R}2, \overline{R}m$ (R even)	-	-	-	$\overline{4}2m$	$\overline{6}m2$	$\overline{4}3m$
$R2$+centre or Rm+centre	$2/m$	mmm	$\overline{3}m$	$4/m\,mm$	$6/m\,mm$	$m3m$

Notes to Table 1.1

1. The word 'centre' refers to the centre of symmetry, $\overline{1}$.

2. The element m is equivalent to $\overline{2}$, with the $\overline{2}$ axis lying normal to the m-plane.

3. R/m means that the axis R is normal to the m-plane; R/m refers to a *single* direction in the point-group symbol (see Table 1.2).

4. Rm means that R lies in the m-plane; $R2$ means that R is normal to 2. Rm and $R2$ each imply *two* directions in the point-group symbol.

Table 1.2   Order and meaning of positions in the point-group symbols

System and point groups;	Position in symbol First	Second	Third	Characteristic symmetry
Triclinic 1, $\bar{1}$	Only one symbol, denoting all directions in the body			None
Monoclinic 2, m, $2/m$	2 or/and $\bar{2}$ along y (y unique)			2 or $\bar{2}$ along y (or z, if z unique)
Orthorhombic 222, $mm2$, mmm	2 or/and along x	2 or/and $\bar{2}$ along y	2 or/and $\bar{2}$ along z	2 or $\bar{2}$ along x, y and z
Tetragonal 4, $\bar{4}$, $4/m$,	4 or/and $\bar{4}$ along z	2 or/and $\bar{2}$ along x and y	-	4 or $\bar{4}$ along z
$4mm$, $\bar{4}\,2m$, 422, $4/m\,mm$	4 or/and $\bar{4}$ along z	2 or/and $\bar{2}$ along x and y	2 or/and $\bar{2}$ along $<110>$	4 or $\bar{4}$ along z
Trigonal 3, $\bar{3}$, 32, $3m$, $\bar{3}\,m$	3 or/and $\bar{3}$ along z	2 or/and $\bar{2}$ along x, y and u	-	3 along z
Hexagonal 6, $\bar{6}$, $6/m$,	6 or/and $\bar{6}$ along z	2 or/and $\bar{2}$ along x, y and u	-	6 or $\bar{6}$ along z
$6mm$, $\bar{6}\,2m$, 622, $6/m\,mm$	6 or/and $\bar{6}$ along z	2 or/and $\bar{2}$ along x, y and u	2 or/and $\bar{2}$ along $<10\bar{1}0>$	6 or $\bar{6}$ along z
Cubic 23, $m3$,	2 or/and $\bar{2}$ along x, y and z	3 or/and $\bar{3}$ along $<111>$	-	3 along $<111>$
432, $\bar{4}\,3m$, $m3m$	4 or/and $\bar{4}$ along x, y and z	3 or/and $\bar{3}$ along $<111>$	2 or/and $\bar{2}$ along $<110>$	3 along $<111>$

Notes to Table 1.2

1. The standard setting of the monoclinic system has y as the unique axis. A $\bar{2}$ axis along y is equivalent to an m-plane normal to y.

2. The trigonal system has been referred here to hexagonal axes. On rhombohedral axes, the first and second positions take the orientations [111] and $<1\,\bar{1}\,0>$, respectively (see Section 1.4.5 for this notation).

3. We space out $4/m\,mm$ and $6/m\,mm$ in order to emphasize the *three* symbol positions for these point groups.

A single symmetry element or an appropriate combination of symmetry elements constitutes a **point group**. It may be defined as *a set of symmetry operations (or elements), the action of which leaves at least one point invariant*, or unmoved. The invariant point is the **origin**, and all symmetry elements pass through it; in some cases, a line or a plane remains invariant under the action of point-group symmetry operations. Since there is, in principle, no limit to the value of R, there is an infinity of point groups; however, we shall restrict our discussion to a more reasonable number of them.

Crystallographic point groups
We focus our attention on the thirty-two crystallographic point groups, as listed in Table 1.1. Not all of them have fully confirmed representatives among chemical entities and, from time to time, we shall introduce certain other, *non-crystallographic* point groups in order to describe the symmetries of some common chemical species. Table 1.1 introduces also the seven crystal **systems**, representing the gross classification of crystals.

Table 1.1 should be studied in conjunction with Table 1.2, which sets out the full meaning of the Hermann-Mauguin (international) symmetry notation used in this book. Table 1.2 indicates also the *minimum* symmetry that characterizes a crystal system. A thorough appreciation of Table 1.2 is germane to our understanding of point groups.

For example, in the orthorhombic system the three positions in the symbol refer to the directions of x, y and z. In the tetragonal system, however, the first position refers to the z direction, the second position to x and y together, and the third to new directions, because the directions of x and y are related by the four-fold rotation that characterizes the tetragonal system. In the orthorhombic system, each of the directions x, y and z is, under orthorhombic symmetry, independent of the other two.

1.3.4 Recognizing point groups
Molecules and crystals may be divided into four symmetry types, dependent upon the presence of a centre of symmetry and one mirror plane or more, or a centre of symmetry alone, or one mirror plane or more but no centre of symmetry, or none of these symmetry elements.

To demonstrate a centre of symmetry, place the given model on a flat surface; then, if the plane through the uppermost atoms (for a chemical species), or the uppermost face (for a crystal), is parallel to the supporting surface, a centre of symmetry is present.

If a mirror plane is present, it divides the model into enantiomorphic (right-hand–left-hand) halves. A correct identification at this stage places the model into one of the four types listed in Table 1.3. The reader may care to examine these rules for a cube or a model of the SF_6 molecule, which shows both a centre of symmetry and mirror planes, and for a tetrahedron or a model of the CH_4 molecule, which shows mirror planes but no centre of symmetry.

Table 1.3 Crystallographic point groups typed by m and/or $\bar{1}$ or neither

Neither m nor $\bar{1}$	Only m	Only $\bar{1}$	Both m and $\bar{1}$
$1, 2, 3, 4, 6, \bar{4}$,	$m, mm2, 3m$,	$\bar{1}, \bar{3}$	$2/m, mmm, \bar{3}m$,
$222, 32, 422$,	$4mm, \bar{4}2m$,		$4/m, 6/m, m3$,
$622, 23, 432$,	$6mm, \bar{6}m2$,		$4/m\,mm, 6/m\,mm$,
	$\bar{6}, \bar{4}3m$		$m3m$

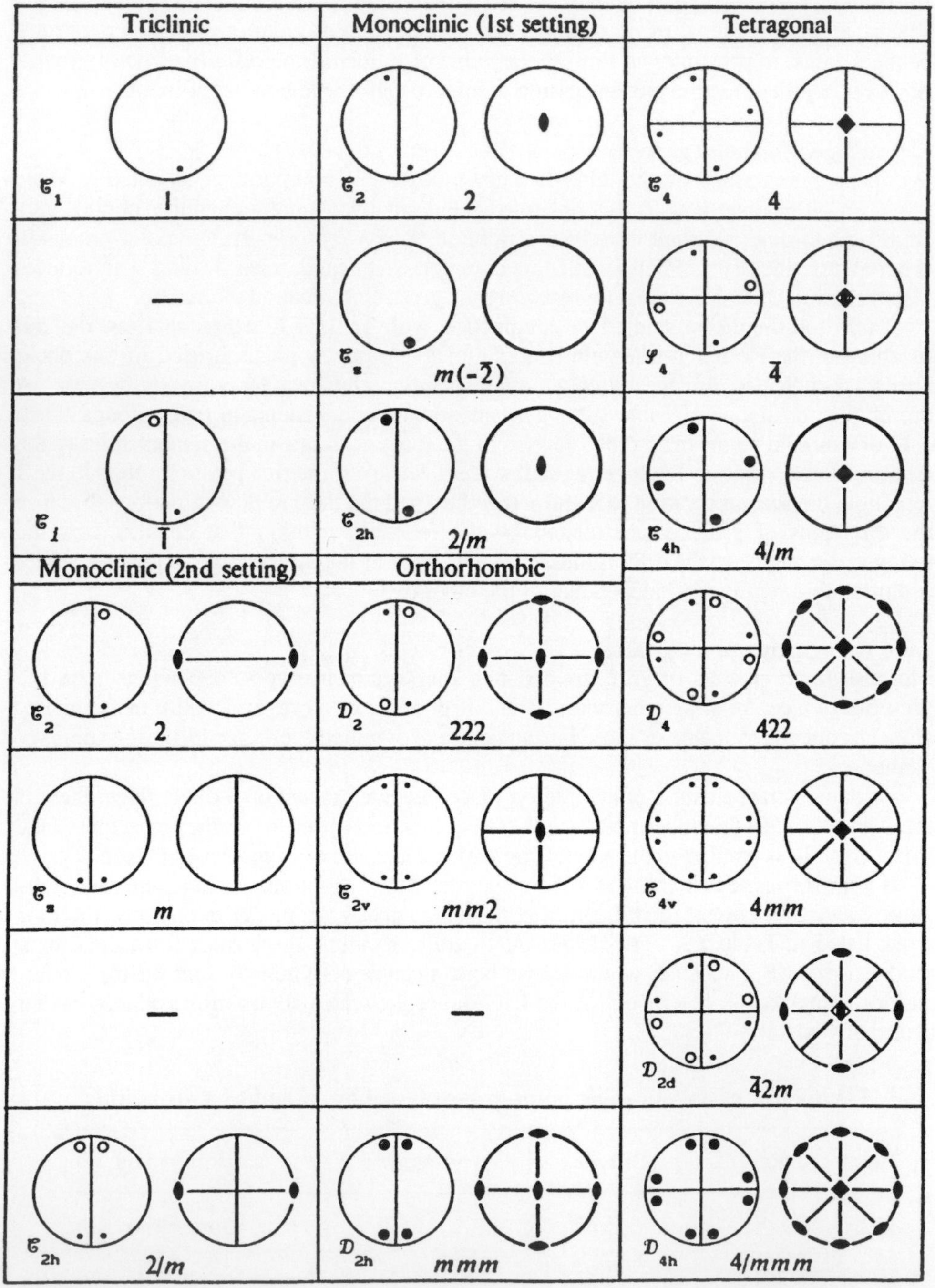

Fig. 1.6. Stereograms of the thirty-two crystallographic point groups; the symbols in the left-hand corners of diagrams show the Schönflies point-group symbols. The y reference axis lies centre to

Trigonal	Hexagonal	Cubic
$\mathcal{C}_3$ 3	$\mathcal{C}_6$ 6	$\mathcal{T}$ 23
—	$\mathcal{C}_{3h}$ $\bar{6}$	—
$\mathcal{C}_6$ $\bar{3}$	$\mathcal{C}_{6h}$ $6/m$	$\mathcal{T}_h$ $m3$
$\mathcal{D}_3$ 32	$\mathcal{D}_6$ 622	$\mathcal{O}$ 432
$\mathcal{C}_{3v}$ $3m$	$\mathcal{C}_{6v}$ $6mm$	—
—	$\mathcal{D}_{3h}$ $\bar{6}m2$	$\mathcal{T}_d$ $\bar{4}3m$
$\mathcal{D}_{3d}$ $\bar{3}m$	$\mathcal{D}_{6h}$ $6/mmm$	$\mathcal{O}_h$ $m3m$

right, z is vertical, except in the triclinic and monoclinic systems, and x lies in the plane of the diagram. The axes coincide with important symmetry elements in the point groups.

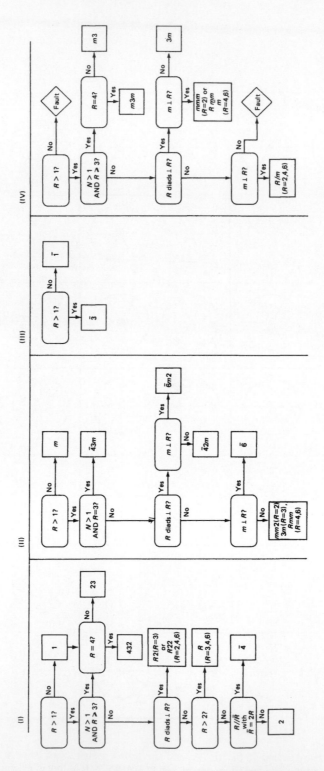

Fig. 1.7. Flow diagram for the point-group recognition program *SYMH*.

Further study of a given model concentrates on the principal rotation axis (the rotation axis of highest degree), the number of them, the presence of mirror planes, two-fold rotation (diad) axes, and so on. The scheme is programmed as *SYMH*, and is accessible as an .EXE file from the Internet address **www.horwood.net/publish.** Figure 1.6 illustrates the crystallographic point groups as stereograms[2,3] showing both the symmetry elements and the general equivalent positions (q.v.). Figure 1.7 is a block diagram of the point-group recognition program. Appendix 1 provides the link between a model and its 'number', as used by the program, and gives also an example or possible example of a chemical species for each point group. For the best results, the model should have been studied carefully along the lines indicated[2] before using the program.

EXAMPLE 1.2. Determine the point group of each of the two models made in Example 1.1. If you are using the program *SYMH*, allocate model number 7 to the cube, and 19 to the tetrahedron. The answers are given at the end of the chapter.

1.4 LATTICES IN ONE, TWO AND THREE DIMENSIONS
A lattice may be defined as *a regular arrangement of (mathematical) points, of infinite extent, such that each point has the same environment as every other point.* This definition means *inter alia* that if we define a **vector** between any two lattice points, then such a vector placed at *any* lattice point will terminate on another lattice point.

The definition applies equally in one dimension, when the lattice is called a **row**, in two dimensions, when it is termed a **net**, and in three dimensions, when it is a **Bravais** lattice. We shall consider lattices in this order of dimensionality.

1.4.1 One-dimensional lattice
The one-dimensional lattice, perhaps the most difficult conceptually, is a row of equidistant points of given spacing, say, *a*. Figure 1.8a shows a row: it could represent the periodicity of the sequence of bricks in Figure 1.8b. Any point may be taken as an origin of the lattice, so that any other point distant vector *r* from the origin is given by

(1.1) $r = Ua$,

where *U* is a positive or negative integer. In one dimension there is only one lattice: a row of points of differing spacing *a′* does not constitute a different *arrangement* of points. The symmetry at each point is that of reflection *m* across the point.

1.4.2 Two-dimensional lattices
If we arrange a series of identical rows of spacing *a* in a regular manner, at a spacing *b* between the rows, we obtain a net, Figure 1.9a, this net could represent the two periodicities of the brick wall, Figure 1.9b. The general net is illustrated by Figure 1.10.

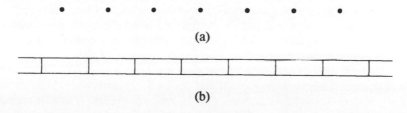

(a)

(b)

Fig. 1.8. One-dimensional lattice. (a) Lattice of spacing *a*. (b) Row of bricks of periodicity *a*.

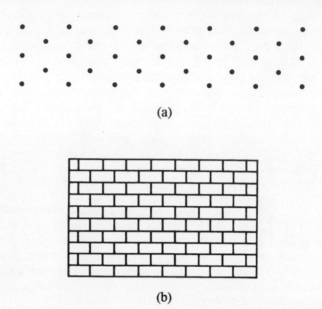

(a)

(b)

Fig.1.9. Two-dimensional lattice. (a) Net of spacings *a* and *b*. (b) Brick wall structure that could be represented by the net in (a), although the spacings are different in their measure.

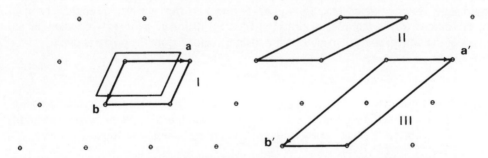

Fig. 1.10. Two-dimensional lattice, with possible unit cells drawn: I and II are primitive *p* and of equal area, whereas III is centred *c* and twice the area of I (or II).

Unit cell

It is convenient to be able to represent a lattice by a unit cell, a portion of the lattice from which the infinite lattice can be generated by repetition in each of the unit cell vectors. We shall use the term **lattice unit cell** for the unit cell of a lattice, **structure unit cell** for the unit cell of a structure, that is, a lattice unit cell with chemical contents, and just **unit cell** when the meaning is unambiguous. The net in Figure 1.10 shows three lattice unit cells. An infinite number of unit cells can be drawn for any lattice. Clearly I and II have the same area, and III has an area of twice that of I or II. A conventional choice of unit cell would be that of I, where the sum *a* + *b* has a minimum value. A further requirement of the unit cell is that the vectors delineating it are chosen along important symmetry directions in the lattice.

Oblique system

The nets can be allocated to two-dimensional systems, something like the crystal systems that we have discussed already. Thus, Figure 1.10 is a net in the **oblique** system: generally, we have $a \neq b$, $\gamma \neq 90°$ or $120°$, where γ is the angle $a{\wedge}b$; a value of $90°$ or $120°$ would lead to higher symmetry in a *lattice*. The symmetry at each point in the oblique net is 2. The **centred** unit cell III in Figure 1.10 does not lead to a second, different oblique lattice. Although we have new dimensions, such that $a' \neq b'$ and $\gamma' \neq 90°$ or $120°$, the unit cell can be re-defined as that in I, which is smaller, and conventional; unit cells I and II are termed **primitive**, *p*.

Rectangular system

When any one of the parameters a, b or γ is specialized in a non-trivial manner ($a' = 2a$ would be trivial), the symmetry at each lattice point is higher than 2. Consider the net in Figure 1.11a: the symmetry at each point is now $2mm$, and the conventional lattice p unit cell is shown by I; it belongs to the **rectangular** system. We have now $a \neq b$ and $\gamma = 90°$. Unit cells II and III have the same area as I, but now a and b do not both lie along the symmetry directions of the lattice.

The centred unit cell in Figure 1.11b reveals a different arrangement of points from that in (a): the p unit cell a', b' of smaller area does not, in isolation, reveal the lattice symmetry clearly. That the rectangular symmetry is present can be inferred from the transformation equations:

(1.2) $$a' = a/2 - b/2 ,$$

(1.3) $$b' = a/2 + b/2 ,$$

whence

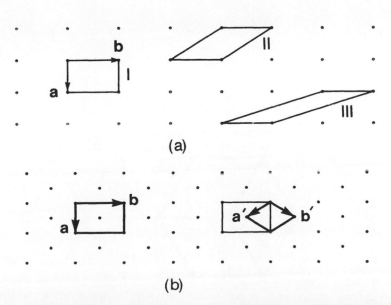

(a)

(b)

Fig. 1.11. Two-dimensional rectangular nets. (a) Conventional p lattice unit cell: a and b are parallel to the m symmetry *lines*; unit cells II and III have the same area as I, but neither is conventional, (b) Unit cell a, b is the conventional centred cell; a', b' is a possible p unit cell, but it is not the conventional choice.

(1.4) $$a' = b' = (a^2/4 + b^2/4)^{1/2}.$$

The angle γ', between a' and b', depends upon the ratio b/a: it is a straightforward matter to show that $\cos(\gamma') = [1/4 - b^2/(4a^2)]/ [1/4 + b^2/(4a^2)]$.

Thus, a parallelogram lattice unit cell in which the sides are equal has symmetry $2mm$. This result is important because one might, at first, identify such a parallelogram with the oblique system, rather than the rectangular system.

Square and hexagonal systems

Further specialization of a and b and γ leads to symmetries $4mm$ (**square**) and $6mm$ (**hexagonal**). We shall discuss these nets by means of a problem, and the five two-dimensional lattices are listed in Table 1.4.

The 'honeycomb' array of points shown in Figure 1.12a is not a two-dimensional lattice: the vector environment of all points is not identical, as shown by the two vectors p_1 and p_2. A true net can be established by centring each hexagon, as shown by Figure 1.12b. However, each centred hexagon, now contains three p hexagonal unit cells of the type specified in Table 1.4.

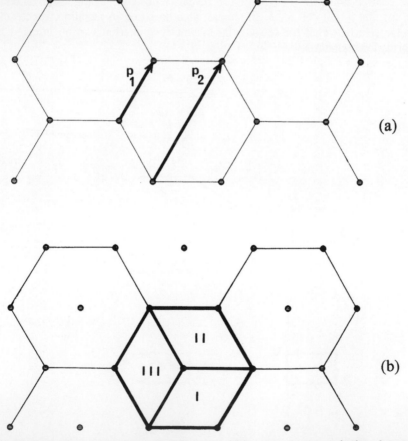

Fig. 1.12. Honeycomb arrangements of points. (a) This array is not a lattice, because the environments of the example vectors p_1 and p_2 are different. (b) This array is a true lattice, but each hexagon now contains *three* hexagonal p unit cells. (Compare with the triply-primitive hexagonal unit cell in Figure 1.35.)

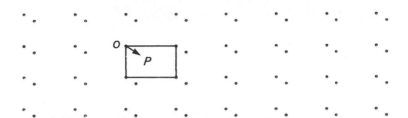

Fig. 1.13. Regular array of points, but not a lattice because all points are not in identical environments. It is best considered as a structure based on a rectangular *p* unit cell with two entities per unit cell, at points such as *O* and *P*.

Table 1.4 Two-dimensional lattices and their conventional lattice unit cells

System	Unit cell/s symbol	Symmetry at each lattice point	Conventional lattice unit cell parameters
Oblique	*p*	2	$a \neq b$; $\gamma \neq 90°$, 120°
Rectangular	*p*, *c*	2*mm*	$a \neq b$; $\gamma = 90°$
Square	*p*	4*mm*	$a = b$; $\gamma = 90°$
Hexagonal	*p*	6*mm*	$a = b$; $\gamma = 120°$

1.4.3 Centring

We have considered lattices for which the conventional lattice unit cells are either primitive or centred. Only a *centring* position is compatible with the definition of a lattice and, therefore, a suitable site for a lattice point. Figure 1.13 shows a regular two-dimensional array of points where the unit cell, origin *O*, contains an additional point, at the end of the vector *OP*. However, the vector *OP* placed at *P* does not terminate on another lattice point.

The whole array of points does not constitute a lattice: it could, however, be a *structure*, based on a rectangular *p* structure unit cell with two entities per unit cell, at the sites *O* and *P*. We shall appreciate this situation more fully when studying space groups.

1.4.4 Three-dimensional, or Bravais, lattices

We consider now a series of identical nets *a*, *b* arranged regularly at a third spacing *c*, whereupon we build up a three-dimensional lattice, a Bravais lattice. Figure 1.14 shows such an array of points in stereoview.

There are fourteen Bravais lattices, distributed unequally among the seven crystal systems listed in Table 1.2. Again, we shall find it convenient to represent each lattice by its conventional lattice unit cell, specified by three non-coplanar vectors *a*, *b* and *c* that are chosen along the reference axes *x*, *y* and *z* respectively. They form a parallelepipedon of minimum volume, subject to the condition that these vectors, like the reference axes, coincide with the important symmetry directions in the lattice. Thus, the conventional lattice unit cell is not always primitive, although a primitive unit cell *can* be selected in any lattice.

We call attention to the general nomenclature that refers to 'primitive' and 'centred' lattices: it is, strictly, a misuse of the term 'lattice'. A lattice is an arrangement of points: to call it primitive or centred implies a certain choice of unit cell. Provided that we are

Fig. 1.14. Steroview of a three-dimensional lattice, obtained by the regular stacking of nets.

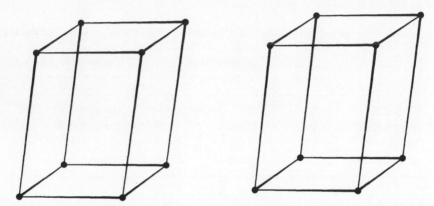

Fig. 1.15. Stereoview of a P unit cell and environs of the triclinic lattice; it may be given the space-group symbol (q.v.) $P\bar{1}$. In this and other three-dimensional illustrations of lattices and structures, unless indicated otherwise, the orientation takes the origin at the bottom, left, rear corner, with $+a$ directed towards the reader, $+b$ directed left to right, and $+c$ directed upwards.

aware of this limitation, no great harm is done by lapsing into the common terminology. We shall consider the Bravais lattices system by system.

Triclinic lattice
Figure 1.15 represents a triclinic lattice. The conventional lattice unit cell is characterized by the conditions
$$a \neq b \neq c; \; \alpha \neq \beta \neq \gamma \neq 90°, 120°.$$
There is only one triclinic lattice, represented by a primitive P unit cell, with the symmetry $\bar{1}$ at each lattice point. Any P lattice unit cell contains one lattice point per unit volume. The volume V of a unit cell may be shown[3] to be given by

(1.5) $V = abc[1 - \cos^2(\alpha) - \cos^2(\beta) - \cos^2(\gamma) + 2\cos(\alpha)\cos(\beta)\cos(\gamma)]^{\frac{1}{2}}$.

Monoclinic lattices
When a Bravais lattice contains symmetry higher than $\bar{1}$ at each point, special relationships exist between the parameters of the conventional unit cell. Thus, in the monoclinic system, we have two-fold symmetry (parallel to b, by convention), with $a \neq b \neq c; \; \alpha = \gamma = 90°; \beta \neq 90°, 120°$. Figure 1.16 shows the monoclinic P lattice.

Fig. 1.16. Stereoview of a P unit cell and environs in a monoclinic lattice; space group $P2/m$.

Table 1.5 Notation for conventional unit cells of Bravais lattices

Centring site/s in unit cell	Unit cell symbol	Miller indices[5] of centred faces	Fractional coordinates of lattice points unique to the unit cell
None	P, R	-	0, 0, 0.
bc faces	A	{100}	0, 0, 0 ; 0, ½, ½.
ca faces	B	{010}	0, 0, 0 ; ½, 0, ½.
ab faces	C	{001)	0, 0, 0 ; ½, ½, 0.
Body	I	-	0, 0, 0 ; ½, ½, ½.
All faces	F	{100}, {010}, {001}	0, 0, 0 ; 0, ½, ½ ; ½, 0, ½ ; ½, ½, 0.

Any centred triclinic unit cell can always be re-selected as a P triclinic cell, while maintaining the characteristic parameters associated with the triclinic unit cell; we need to consider the same question in the monoclinic system.

In three dimensions, a unit cell may be centred on any pair of opposite faces, on all faces or at the geometrical centre (body-centre) of the unit parallelepipedon.   These possible arrangements lead to the notation for unit cells as listed in Table 1.5.

The **fractional coordinates** in Table 1.5 are defined such that a fractional coordinate x is given by X/a, where X is the coordinate in absolute measure and a is the unit cell length, parallel to x, in the same measure; thus, they are independent of the unit cell dimensions.

A monoclinic unit cell centred on the C faces is shown in Figure 1.17. An A-centred monoclinic unit cell does not represent a new arrangement, because it becomes a C unit cell by the transformation

(1.6)
$$\mathbf{a}_C = \mathbf{c}_A,$$
$$\mathbf{b}_C = -\mathbf{b}_A,$$
$$\mathbf{c}_C = \mathbf{a}_A.$$

The symmetry axis is still parallel to b, but the sign change is needed to preserve a right-handed nature of the axes.

Centring on the B faces is illustrated by Figure 1.18; two adjacent unit cells are shown. It is evident that a smaller, conventional P monoclinic unit cell may be drawn, defined by the transformation equations

(1.7)
$$\begin{aligned}
\boldsymbol{a'} &= \boldsymbol{a}\,, \\
\boldsymbol{b'} &= \boldsymbol{b}\,, \\
\boldsymbol{c'} &= -\boldsymbol{a}/2 + \boldsymbol{c}/2\,.
\end{aligned}$$

The new P cell has half the volume of the B-centred unit cell. If two or more unit cells are outlined within one and the same lattice, the ratio of the volumes of the unit cells is given by the direct ratio of the numbers of lattice points unique to them; thus, $V_P = V_B/2$.

In order to show the principle of the sharing of lattice points by unit cells, Figure 1.19 illustrates eight adjacent monoclinic unit cells: it is evident that a lattice point at the corner of a unit cell is shared equally by eight unit cells. The sharing principles are set out fully in Table 1.6.

If we transform the monoclinic C unit cell to P, Figure 1.20, we obtain a unit cell in which $a \neq b \neq c$; $\alpha = 90°$; $\beta \neq \gamma \neq 90°$, $120°$. This result points to the fact that monoclinic C represents an arrangement of points different from monoclinic P, and so constitutes a new lattice.

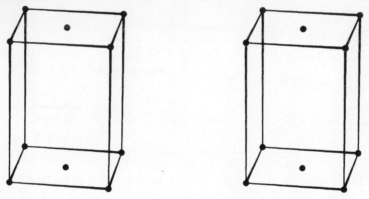

Fig. 1.17. Stereoview of a C unit cell and environs in a monoclinic lattice ($C2/m$).

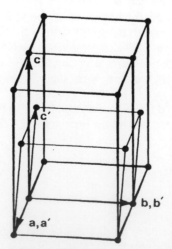

Fig.1.18. Monoclinic lattice, with two adjacent B unit cells; the equivalent P unit cell is shown.

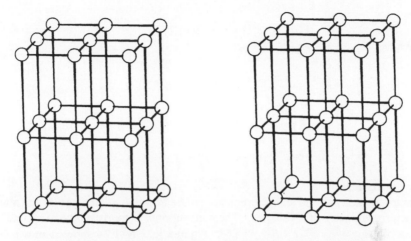

Fig. 1.19. Stereoview of eight adjacent monoclinic P unit cells.

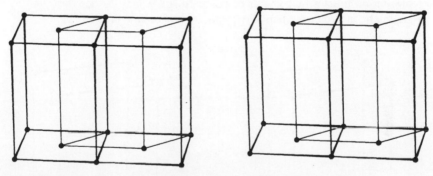

Fig. 1.20. Stereoview of a two monoclinic C unit cells; the P unit cell outlined is not a conventional choice for this lattice.

Table 1.6 Sharing of lattice points among the Bravais unit cells

Lattice point site	Sharing principle	Contribution to unit cell
Corner	8 adjacent cells	1/8
Mid-point of edge	4 adjacent cells	1/4
Face centre	2 adjacent cells	1/2
Body centre	1 unit cell	1

Transformations may be involved in practical situations. Using equation (1.7), we can evaluate the constants for the transformed unit cell by simple vector methods[3]. We need to find expressions for c' and β' in terms of the original cell parameters. Thus:

(1.8) $$c' \cdot c' = (-a/2 + c/2) \cdot (-a/2 + c/2) ,$$

whence

$$c'^2 = a^2/4 + c^2/4 - ac \cos(c {^\wedge} a)/2 ,$$

so that

$$c' = [a^2/4 + c^2/4 - ac \cos(\beta)/2]^{\frac{1}{2}}.$$

For β', we have

(1.9) $a' \cdot c' = a'c' \cos(\beta')$,

whence

$$\cos(\beta') = a' \cdot (-a/2 + c/2)/a'c' = [-a^2/2 + ac \cos(\beta)/2]/ac' ,$$

so that

$$\beta' = \cos^{-1}\{[-a + c \cos(\beta)]/(2c')\} .$$

Orthorhombic lattices

The symmetry at each point in an orthorhombic lattice is *mmm*; a lattice point site always has the highest symmetry of its system. The conventional orthorhombic lattice unit cell has the conditions $a \neq b \neq c$; $\alpha = \beta = \gamma = 90°$. Figures 1.21 to 1.24 show stereoviews of the orthorhombic P, C, I and F lattice unit cells. These unit cells are distinct, and they correspond to four different lattices with orthorhombic symmetry. That these four unit cells do represent different arrangements of points in space may be demonstrated by the following argument, already used implicitly with the monoclinic lattices.

Fig 1.21. Stereoview of the P unit cell and environs of an orthorhombic lattice (*Pmmm*).

 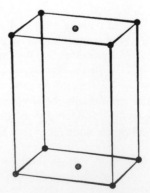

Fig 1.22. Stereoview of the C unit cell and environs of an orthorhombic lattice (*Cmmm*).

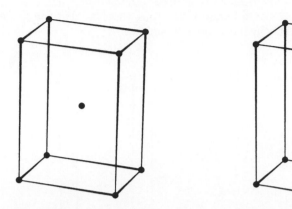

Fig 1.23. Stereoview of the *I* unit cell and environs of an orthorhombic lattice (*Immm*).

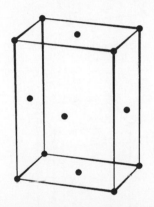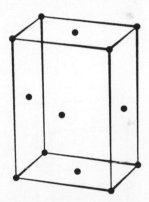

Fig 1.24. Stereoview of the *F* unit cell and environs of an orthorhombic lattice (*Fmmm*).

After centring a *P* unit cell in a given system, we consider the following questions:
(a) Does the centred unit cell represent a lattice?
(b) If it is a true lattice, is the symmetry different from that of the *P* unit cell?
(c) If the symmetry is unchanged, is the lattice different in type (arrangement of points) from the lattice or lattices already determined for the given system? It may be judged by reducing the unit cell to one of a lower degree of centring, or to primitive, and then comparing the relationships of the reduced unit-cell parameters with those characteristic for the system. This last test is equivalent to asking if the unit cell has been chosen correctly, according to convention.

Tetragonal lattices
There are two tetragonal lattices, symmetry 4/*m mm* at each point, represented by *P* and *I* unit cells, Figures 1.25 and 1.26. For each of them, we have the conditions $a = b = c$; $\alpha = \beta = \gamma = 90°$. It is straightforward to show that $C \equiv P$, Figure 1.27.

If the *B* faces are centred, the characteristic tetragonal symmetry is lost; the unit cell is now orthorhombic, even though $a = b$. The tetragonal symmetry is restored, apparently, by centring the *A* faces as well. However, this cell does not represent a lattice, because the two vectors p_1p_1' and p_2p_2', illustrated in Figure 1.28, do not have identical environments.

Centring the C faces in addition to A and B produces a tetragonal F unit cell, but this is equivalent to the I unit cell shown in Figure 1.26. Thus, there are only two tetragonal lattices, represented conventionally by the symbols P and I.

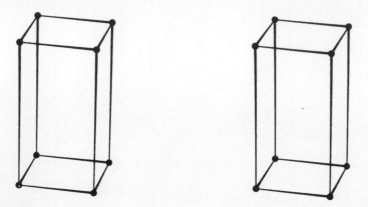

Fig 1.25. Stereoview of the P unit cell and environs of a tetragonal lattice ($P4/m\ mm$).

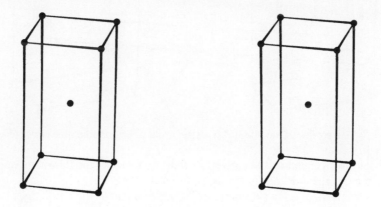

Fig 1.26. Stereoview of the I unit cell and environs of a tetragonal lattice ($I4/m\ mm$).

Fig 1.27. Tetragonal lattice, showing the equivalence of P and C descriptors.

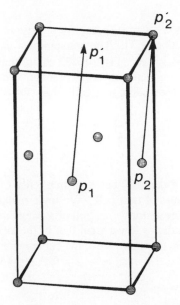

Fig. 1.28. A 'tetragonal' lattice unit cell centred on the A and B faces does not represent a lattice.

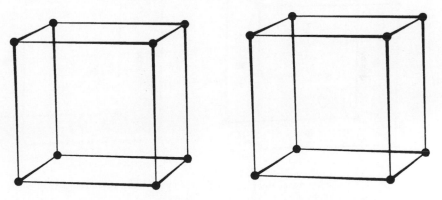

Fig 1.29. Stereoview of the P unit cell and environs of a cubic lattice ($Fm3m$).

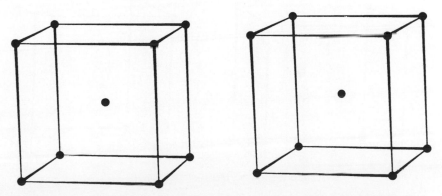

Fig 1.30. Stereoview of the I unit cell and environs of a cubic lattice ($Im3m$).

Cubic lattices

There are three cubic lattices, P, I and F, that are consistent with the symmetry $m3m$ at each lattice point, Figures 1.29 to 1.31. The conventional unit cells correspond with the parameters $a = b = c$; $\alpha = \beta = \gamma = 90°$. Figure 1.32 shows that centring the A faces of a P cubic unit cell reduces the symmetry to orthorhombic, because the symmetry at each lattice point is now *mmm*. If the I cubic unit cell is transformed by the equations

(1.10)
$$a' = a ,$$
$$b' = b ,$$
$$c' = a + c ,$$

the new lattice unit cell is A-centred, but it is implicitly of cubic symmetry: a lattice remains invariant under change of unit cell. We must remember that the unit cell of a lattice, or a structure, is a *representative portion* of a whole array, although the unit cell in isolation may have a symmetry lower than that of the lattice from which it has been derived. In the particular case of Figure 1.33, the cubic symmetry is inherent in the non-trivial relationships $a' = b' = a, c' = a\sqrt{2}$; $\alpha' = \gamma' = 90°$, $\beta = 45°$.

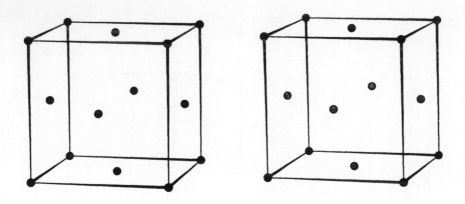

Fig 1.31. Stereoview of the F unit cell and environs of a cubic lattice ($Fm3m$).

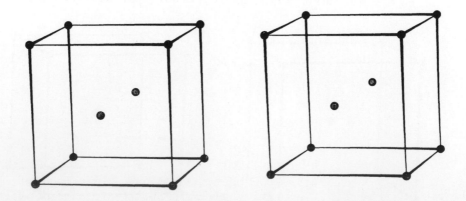

Fig. 1.32. Stereoview of an A-centred 'cubic' unit cell that is really orthorhombic; the symmetry, originally $m3m$, has been reduced to *mmm* at each point by the A-centring.

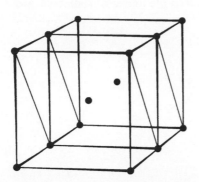
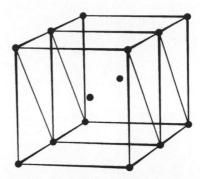

Fig 1.33. For the unit cell outlined by thin lines $a = b \neq c$, $\alpha = \gamma = 90°$, $\beta = 45°$. The symmetry at each lattice point remains $m3m$, but an A-centred unit cell is unconventional for the cubic system; the conventional unit cell is I (thick lines).

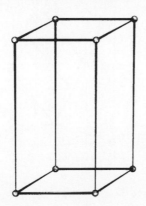
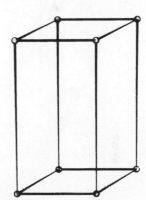

Fig. 1.34. Steroview of the P unit cell and environs of a hexagonal lattice ($P6/m\ mm$).

Hexagonal lattice

There is a single hexagonal lattice, Figure 1.34, and the P unit cell has the parameters $a = b \neq c$, $\alpha = \beta = 90°$, $\gamma = 120°$; the symmetry at each lattice point is $6/m\ mm$. The six-fold symmetry becomes apparent as the unit cells are stacked together.

Trigonal lattices

The two-dimensional hexagonal unit cell, in which $a = b$ and $\gamma = 120°$, is compatible with both three-fold and six-fold symmetry. Thus, the hexagonal P unit cell (Figure 1.34) may be used for certain lattices of trigonal symmetry. However, if the lattice has three-fold rotation axes, parallel to z, and passing through the points 2/3, 1/3 and 1/3, 2/3 in the a,b plane, a triply-primitive hexagonal unit cell arises, Figure 1.35a. Hence, some trigonal crystals will have a true P unit cell, whereas others will have a triply-primitive hexagonal unit cell.

The triply-primitive unit cell can be transformed to a primitive rhombohedral (R) unit cell, Figure 1.35a, in which $a = b = c$, $\alpha = \beta = \gamma \neq 90°$ and $< 120°$. The unique angle is chosen to be less than 120°, rather than its supplement, so that the volume of the unit cell from (1.5) calculates as positive. In the R unit cell, the three-fold axis is along [111], and the cell may be thought of as a P cubic unit cell extended along one of its three-fold axes.

The transformation from the triple hexagonal unit cell to the (obverse) rhombohedral unit cell is given by the equations

(1.11)
$$a_R = 2a_H/3 + b_H/3 + c_H/3 ,$$
$$b_R = -a_H/3 + b_H/3 + c_H/3 ,$$
$$c_R = -a_H/3 - 2b_H/3 + c_H/3 ,$$

and the inverse transformation is given by

(1.12)
$$a_H = a_R - b_R ,$$
$$b_H = b_R - c_R ,$$
$$c_H = a_R + b_R + c_R .$$

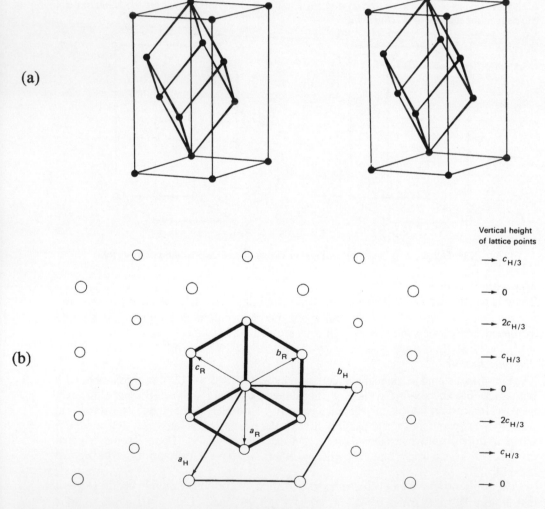

(a)

(b)

Vertical height
of lattice points

$\longrightarrow c_{H/3}$

$\longrightarrow 0$

$\longrightarrow 2c_{H/3}$

$\longrightarrow c_{H/3}$

$\longrightarrow 0$

$\longrightarrow 2c_{H/3}$

$\longrightarrow c_{H/3}$

$\longrightarrow 0$

Fig 1.35. (a) Stereoview of the *R* (primitive) rhombohedral unit cell in the *obverse* setting with respect to the triply-primitive hexagonal unit cell. (b) Projection of (a) along [111] in the rhombohedral unit cell; the fractions relate to vertical heights, along c_H.

The alternative, *reverse*, setting of the R unit cell is obtained by rotating the rhombohedral unit cell, Figure 1.35b, clockwise about c_H by 60°.

Table 1.7 presents a summary of the fourteen Bravais lattices that we have described in this section, with the axial relationships in the conventional unit cell.

1.4.5 Directions

Lattice geometry is governed through the three basic vector translations *a*, *b* and *c*. It follows from the definition of a lattice that any point may be reached from any other given point by performing the basic translations, or multiples thereof, always in the directions of *a*, *b* and *c*.

Any lattice point may be taken as the origin, and a vector *r* from the origin to any lattice point is given by

(1.13) $$r = Ua + Vb + Wc,$$

where U, V and W are positive or negative integers or zero, the coordinates of the given lattice point. Any point x, y, z *within* the unit cell is distant d from the origin, where

Table 1.7 The Bravais lattices

Crystal system	Unit cell symbol/s	Axial relationships in conventional unit cell	Symmetry at each lattice point
Triclinic	*P*	$a \neq b \neq c$ $\alpha \neq \beta \neq \gamma \neq 90°, 120°$	$\bar{1}$
Monoclinic	*P, C*	$a \neq b \neq c$ $\alpha \neq \beta \neq \gamma \neq 90°, 120°$	$2/m$
Orthorhombic	*P, C, I, F*	$a \neq b \neq c$ $\alpha = \beta = \gamma = 90°$	*mmm*
Tetragonal	*P, I*	$a = b \neq c$ $\alpha = \beta = \gamma = 90°$	$4/m\ mm$
Cubic	*P, I, F*	$a = b = c$ $\alpha = \beta = \gamma = 90°$	$m3m$
Hexagonal	*P*	$a = b \neq c$ $\alpha = \beta = 90°, \gamma = 120°$	$6/m\ mm$
Trigonal (Hexagonal axes)	*P*	$a = b \neq c$ $\alpha = \beta = 90°, \gamma = 120°$	$\bar{3}m$
Trigonal (Rhombohedral axes)	*R*	$a = b = c$ $\alpha = \beta = \gamma \neq 90°, <120°$	$\bar{3}m$

(1.14) $d = x\boldsymbol{a} + y\boldsymbol{b} + z\boldsymbol{c}$.

The line joining the origin to the lattice points U, V, W; $2U$, $2V$, $2W$; ... nU, nV, nW; defines a **direction**, or directed line, $[UVW]$ in the lattice. A set of directions $<UVW>$ related by symmetry defines a **form** of directions. Thus, $[123]$ under symmetry $2/m$ is related to $[1\bar{2}3]$, $[\bar{1}2\bar{3}]$ and $[\bar{1}\,\bar{2}\,\bar{3}]$, and the four directions comprise the form $<123>$. A numerical value for r or d may be obtained by following (1.8).

1.4.6 Rotation symmetry of crystals

We show how a crystal, based on a Bravais lattice, can possess symmetry consistent with the degrees of rotation 1, 2, 3, 4 and 6 only.

In Figure 1.36, A and B represents two adjacent points in any row of a lattice; their repeat distance along that row is t. An R-fold rotation axis will be considered to act at each lattice point, in a direction normal to a plane containing A and B, for example, the plane of the diagram.

An anticlockwise rotation of ϕ about the axis through A maps B on to B', and a similar but clockwise rotation about the axis through B maps A on to A'. The lines $A'S$ and $B'T$ are drawn normal to the line through A and B to meet it in S and T, respectively.

Triangles $AB'T$ and $BA'S$ are congruent, so that $A'B'$ is parallel to AB. In a lattice, any two points in a row must be separated by an integral multiple of the repeat distance in that row. Thus, we have

$$A'B' = mt ,$$

where m is an integer. Furthermore,

$$A'B' = t - (AT + BS) = t - 2t \cos(\phi) ,$$

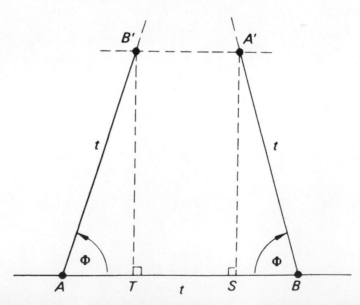

Fig. 1.36. Illustration for the possible rotational symmetries for crystals based on a Bravais lattice. Rotation axes in the lattice have been set normal to the plane of the diagram, passing through points such as A, B, B' and B'.

whence

$$m = 1 - 2\cos(\phi) \,,$$

or

$$\cos(\phi) = (1 - m)/2 = M/2 \,,$$

where M is another integer. Since $|\cos(\phi)| \leq 1$, we find the permissible results as follow:

ϕ/deg	0(360)	180	120	90	60
M	2	−2	−1	0	1
R	1	2	3	4	6

1.5 SPACE GROUPS

In the building up of crystal structures, chemical entities, which may or may not themselves possess symmetry, are associated in a regular manner with the points of one of the Bravais lattices. This extended array of atoms and molecules in a lattice base brings us to the study of **space groups**. Space groups have been treated in detail elsewhere[2–4] and the reader's attention is drawn to these standard works. Our treatment here will be relatively brief, but is needed for completeness.

A space group may be defined as *an infinite set of symmetry elements, the action with respect to which maps the pattern to which it refers on to itself.* It follows that the pattern, whether in one, two or three dimensions, after the action of any symmetry element in the set, is indistinguishable from its condition before the action. We shall consider a small number of space groups, from the monoclinic, orthorhombic, tetragonal, hexagonal and cubic crystal systems.

1.5.1 Monoclinic space groups

A space group can be formed by taking a chemical moiety and arranging it in an identical orientation at or around each point of a Bravais lattice. Space groups that are derived in this way are termed **symmorphic** space groups, and an example is the space group represented by the symbol $P2/m$. From our discussion so far, we know that we are concerned with a monoclinic space group, based on a P unit cell, with 2 parallel to y and an m-plane parallel to the x,z plane. The two-fold axis intersects the mirror plane in a centre of symmetry $\bar{1}$.

We can illustrate this space group by a drawing that shows the symmetry elements and the manner in which they repeat a given moiety in and around the unit cell. Figure 1.37 shows space group $P2/m$: by convention, $+x$ runs from top to bottom, $+y$ from left to right and $+z$ is directed upwards from the plane of the diagram. The right-hand diagram shows the symmetry elements: arrows indicate the two-fold axes, the thick lines show the m-planes and the small circles the centres of symmetry; in this group, the latter sites have the symmetry $2/m$.

The left-hand diagram shows how a general point x, y, z is repeated by the symmetry to give four *general equivalent positions*, the sites labelled o in the Wyckoff notation. *Special* equivalent positions exist at sites of point-group (*non-translational*) symmetry. They have symmetry m for the Wyckoff n and m positions, symmetry 2 for l, k, j and i, and so on. Figure 1.16 could be regarded as either a monoclinic P lattice, or as a structure of space group $P2/m$ with a moiety of symmetry $2/m$ at Wyckoff positions a. In determining the sets of special equivalent positions, there are two conditions to fulfil:

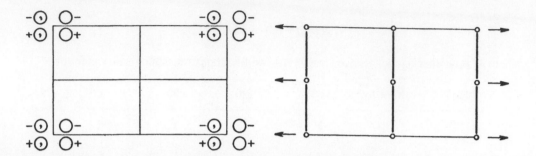

Origin at centre ($2/m$); unique direction b.

Number of positions	Wyckoff notation	Point symmetry	Coordinates of equivalent positions
4	o	1	$x,y,z ;\ \bar{x} ,\ \bar{y} ,\ \bar{z} ;\ x,\ \bar{y} ,z ;\ \bar{x} ,y,\ \bar{z}$.
2	n	m	$x, \tfrac{1}{2}, z ;\ \bar{x} , \tfrac{1}{2},\ \bar{z}$.
2	m	m	$x, 0,\ z ;\ \bar{x} , 0 ,\ \bar{z}$.
2	l	2	$\tfrac{1}{2}, y, \tfrac{1}{2} ;\ \tfrac{1}{2},\ \bar{y} , \tfrac{1}{2}$.
2	k	2	$0, y, \tfrac{1}{2} ;\ 0,\ \bar{y} , \tfrac{1}{2}$.
2	j	2	$\tfrac{1}{2}, y, 0 ;\ \tfrac{1}{2},\ \bar{y} , 0$.
2	i	2	$0, y, 0 ;\ 0,\ \bar{y} , 0$.
1	h	$2/m$	$\tfrac{1}{2},\ \tfrac{1}{2},\ \tfrac{1}{2}$.
1	g	$2/m$	$\tfrac{1}{2},\ 0,\ \tfrac{1}{2}$.
1	f	$2/m$	$0,\ \tfrac{1}{2},\ \tfrac{1}{2}$.
1	e	$2/m$	$\tfrac{1}{2},\ \tfrac{1}{2},\ 0$.
1	d	$2/m$	$\tfrac{1}{2},\ 0,\ \tfrac{1}{2}$.
1	c	$2/m$	$0,\ 0,\ \tfrac{1}{2}$.
1	b	$2/m$	$0,\ \tfrac{1}{2},\ 0$.
1	a	$2/m$	$0,\ 0,\ 0$.

Fig. 1.37. Drawing to show the general equivalent positions and symmetry elements in space group *P2/m*. The origin is on a symmetry centre ($2/m$); general and special equivalent positions are listed in their sets. The arrows indicate two-fold axes, the thick line the m-planes and the small circles the centres of symmetry. The ± signs refer to the third coordinate, z. The Wyckoff notation $a, b, c, \ldots$ is simply a way of referring collectively to sets of equivalent positions.

(1) The sets must be true sub-sets of the general equivalent positions: thus, the site $0, y, 0$ forms a set with $0,\ \bar{y},\ 0$, as can be confirmed by setting $x = z = 0$ in the general equivalent positions.

(2) There are always symmetry elements, belonging to the unit cell, above the plane of the diagram; for example, 2 at $0, y, \tfrac{1}{2}$, relating, say, x, y, z to $\bar{x}, y, 1 - z$.

An important feature of space groups is the existence of translational symmetry elements, other than lattice centring itself, namely, **screw-axes** and **glide-planes**. In some symmorphic groups they arise naturally: thus, *C2/m* exhibits a-glide-planes, parallel to the *m*-planes[4], at $y = \frac{1}{4}$ and $\frac{3}{4}$. They relate points such as x, y, z and $\frac{1}{2} + x, \frac{1}{2} - y, z$: the operation comprises two movements, a reflection across the *a*-glide-plane, transiently to $(x, \frac{1}{2} - y, z)$, followed by a translation of $\frac{1}{2}$ in the direction of the *a* unit cell translation to give $\frac{1}{2} + x, \frac{1}{2} - y, z$, the two movements making up the single glide-symmetry operation. Not all space groups can be built up by combining a point group with a lattice translation; in fact, only 73 can be produced in this way. The remaining 157 (230 − 73) are termed **non-symmorphic** space groups: one or more translational symmetry elements, other than lattice translations, must be defined at the outset, and one example of a non-symmorphic space group is *C*2/*c*, Figure 1.38.

The *C* centring gives rise to 2_1 screw-axes (interleaving the two-fold axes) which combine two-fold rotation with a translation of $\frac{1}{2}$, along the *b* direction; the screw-axes are symbolized by half-arrows on the diagram. In combination with the *c*-glide planes, the centring introduces additional symmetry elements *within* each unit cell: they are *n*-glide-planes (parallel to and interleaving the *c*-glide-planes) with translational components of $(a + c)/2$. The reader should identify each symmetry element and the general equivalent position it produces starting from x, y, z. One way of ensuring that all symmetry elements have been included on a diagram is to check that any general equivalent position can be reached from any other by a *single* symmetry operation.

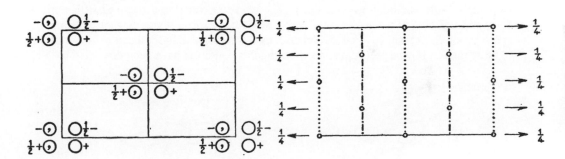

Origin at $\bar{1}$, on *c*-glide-plane; unique direction *b*.

Number of positions	Wyckoff notation	Point symmetry	Coordinates of equivalent positions
			$(0, 0, 0;\ \frac{1}{2}, \frac{1}{2}, 0)+$
8	*f*	1	$x, y, z;\ \bar{x}, \bar{y}, \bar{z};\ x, \bar{y}, \frac{1}{2}+z;\ \bar{x}, y, \frac{1}{2}-z.$
4	*e*	2	$0, y, \frac{1}{4};\ 0, \bar{y}, \frac{3}{4}.$
4	*d*	$\bar{1}$	$\frac{1}{4}, \frac{1}{4}, \frac{1}{2};\ \frac{3}{4}, \frac{1}{4}, 0.$
4	*c*	$\bar{1}$	$\frac{1}{4}, \frac{1}{4}, 0;\ \frac{3}{4}, \frac{1}{4}, \frac{1}{2}.$
4	*b*	$\bar{1}$	$0, \frac{1}{2}, 0;\ 0, \frac{1}{2}, \frac{1}{2}.$
4	*a*	$\bar{1}$	$0, 0, 0;\ 0, 0, \frac{1}{2}.$

Fig. 1.38. Space group *C*2/*c*. The origin is on a symmetry centre $\bar{1}$, and the general and special equivalent positions are listed in their sets, headed by the *C* centring conditions. The half-arrows indicate 2_1 screw-axes, dotted lines are *c*-glide-planes, and dot–dash lines are *n*-glide-planes. Fractions indicate heights in *z* as fractions of *c*.

The translations of the *C* unit cell, shown by the expression (0, 0, 0; ½, ½, 0)+ placed above the list of coordinates, are to be added to each coordinate in the list. Thus, the *e* positions, *in extenso*, read 0, *y*, ¼ ; 0, $\bar{y}$, ¾; ½, ½ + *y*, ¼; ½, ½ − *y*, ¾.

Diiodo(N,N,N′,N′-tetramethylethylenediamine)zinc(II) has the structural formula $(CH_3)_2NCH_2CH_2N(CH_3)_2ZnI_2$ and crystallizes in space group *C2/c* with four molecules in the structure unit cell. Figure 1.39a is a stereoview of the molecular structure, and Figure 1.39b is a stereoview of the unit cell showing, for convenience, the positions of the iodine and zinc atoms.

It is evident that the structure is based on the monoclinic lattice in Figure 1.17. Since there are four molecules in the unit cell and because they are not centro-symmetric, they lie athwart the two-fold axes, with the zinc atoms on these axes. The reader should examine Figures 1.38 and 1.39 together, with a view to identifying the symmetry elements of the space group in the structure unit cell.

1.5.2 Orthorhombic, tetragonal, hexagonal and cubic space groups

For the orthorhombic system, we have chosen the space group *Pnma*. We note first how the Hermann-Mauguin notation reveals all that we need to know about the relative positions of the symmetry elements. If we remove the lattice symbol *P* and convert the translational symmetry to the corresponding non-translational symmetry elements, we obtain the point group *mmm*, which we know to be orthorhombic. By referring to the important Table 1.2, it follows that the *n*-glide-plane is normal to *x*, the *m*-plane is normal to *y*, and the *a*-glide-plane is normal to *z*. Since point group *mmm* is centrosymmetric, space group *Pnma* will be centrosymmetric, too. It is convenient for a centre of symmetry to be at the origin, and procedures for ensuring this condition have been discussed by the

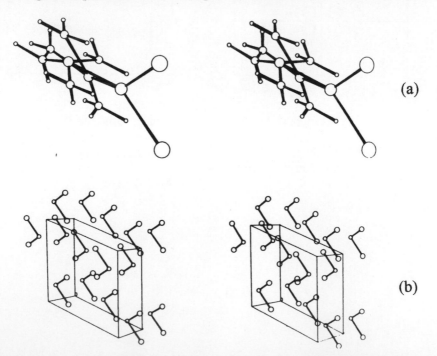

(a)

(b)

Fig. 1.39 Stereoviews of diiodo(N,N,N′,N′-tetramethylethylenediamine)zinc(II). (a) Structural formula, in the same orientation as in (b); the circles in order of decreasing size represent the I, Zn, N, C and H species. (b) Unit cell, showing for convenience, only the I and Zn species.

author elsewhere[2,3]. Figure 1.40 illustrates the general and special equivalent positions for the space group *Pnma*.

Dirhodium boron Rh_2B crystallizes in this space group. Its unit cell dimensions are $a = 0.542$ nm, $b = 0.398$ nm, $c = 0.744$ nm; the density of the crystals is 8960 kg m^{-3}. Thus, the number of formula-entities per unit cell is $[8960 \times (0.542 \times 0.398 \times 0.744 \times 10^{-27}]/(216.63 \times 1.6605 \times 10^{-27}) = 3.998$, or 4 to the nearest integer.

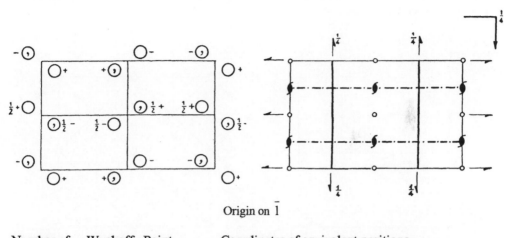

Origin on $\bar{1}$

Number of positions	Wyckoff notation	Point symmetry	Coordinates of equivalent positions
8	d	1	$\pm(x, y, z\,;\, \tfrac{1}{2}+x, \tfrac{1}{2}-y, \tfrac{1}{2}-z\,;\, \bar{x}, \tfrac{1}{2}+y, \bar{z}\,;\, \tfrac{1}{2}-x, \bar{y}, \tfrac{1}{2}+z)$
4	c	m	$x, \tfrac{1}{4}, z\,;\, \bar{x}, \tfrac{3}{4}, \bar{z}\,;\, \tfrac{1}{2}-x, \tfrac{3}{4}, \tfrac{1}{2}+z\,;\, \tfrac{1}{2}+x, \tfrac{1}{4}, \tfrac{1}{2}-z$.
4	b	$\bar{1}$	$0, 0, \tfrac{1}{2}\,;\, 0, \tfrac{1}{2}, \tfrac{1}{2}\,;\, \tfrac{1}{2}, 0, \tfrac{1}{2}\,;\, \tfrac{1}{2}, \tfrac{1}{2}, 0$.
4	a	$\bar{1}$	$0, 0, 0\,;\, 0, \tfrac{1}{2}, 0\,;\, \tfrac{1}{2}, 0, \tfrac{1}{2}\,;\, \tfrac{1}{2}, \tfrac{1}{2}, \tfrac{1}{2}$.

Fig. 1.40. Drawings to show the general equivalent positions and symmetry elements in space group *Pnma*. The 'two-tail' ellipses represent 2_1 screw-axes parallel to z, and the right-angled arrow indicates a-glide planes at $\tfrac{1}{4}$ and $\tfrac{3}{4}$ (along c).

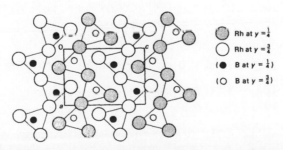

Fig. 1.41. Unit cell and environs of the structure of RhB_2, as seen along the direction of the y axis. The large circles represent Rh atoms on the m-planes at $y = \tfrac{1}{4}$ and $\tfrac{3}{4}$, and the small circles represent B atoms on the m-planes at $y = \tfrac{1}{4}$ and $\tfrac{3}{4}$. (R Mooney and A J E Welch, *Acta Cryst.* **7**, 49, 1954.)

The eight rhodium atoms can lie in a set of general positions or in two sets of special positions. With the rhodium atoms in general positions, the separation of atoms related by the m-plane would be $\frac{1}{2} - 2y$. Thus, the maximum separation in the y direction is less than 0.2 nm, which is insufficient for two nearest-neighbour rhodium atoms ($r \approx 0.12$ nm). Hence, the rhodium atoms must be placed in special positions, on sites of either m or $\bar{1}$ symmetry. By similar reasoning, the positions on $\bar{1}$ may be eliminated, so that two sets of c positions are occupied. Figure 1.41 is a diagrammatic illustration of the structure, as seen along the y axis.

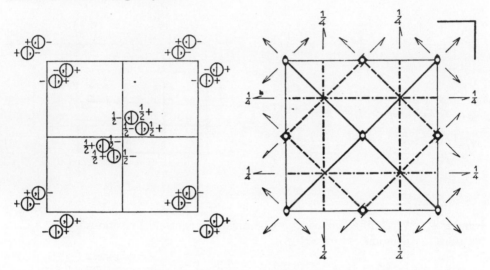

Origin at centre (mmm)

Number of positions	Wyckoff notation	Point symmetry	Coordinates of equivalent positions
8	h	2	$0\,\frac{1}{2},z$; $0,\frac{1}{2},\bar{z}$; $0,\frac{1}{2},\frac{1}{2}+z$; $0,\frac{1}{2},\frac{1}{2}+z$; $\frac{1}{2},0,z$; $\frac{1}{2},0,\bar{z}$; $\frac{1}{2},0,\frac{1}{2}+z$; $\frac{1}{2},0,\frac{1}{2}+z$.
4	g	$mm2$	$x,\bar{x},0$; $\bar{x},x,0$; $\frac{1}{2}+x,\frac{1}{2}+x,\frac{1}{2}$; $\frac{1}{2}-x,\frac{1}{2}-x,\frac{1}{2}$.
4	f	$mm2$	$x,x,0$; $\bar{x},\bar{x},0$; $\frac{1}{2}+x,\frac{1}{2}-x,\frac{1}{2}$; $\frac{1}{2}-x,\frac{1}{2}+x,\frac{1}{2}$.
4	e	$mm2$	$0,0,z$; $0,0,\bar{z}$; $\frac{1}{2},\frac{1}{2},\frac{1}{2}+z$; $\frac{1}{2},\frac{1}{2},\frac{1}{2}-z$.
4	d	$\bar{4}$	$0,\frac{1}{2},\frac{1}{4}$; $\frac{1}{2},0,\frac{1}{4}$; $0,\frac{1}{2},\frac{3}{4}$; $\frac{1}{2},0,\frac{3}{4}$.
4	c	$2/m$	$0,\frac{1}{2},0$; $\frac{1}{2},0,0$; $0,\frac{1}{2},\frac{1}{2}$; $\frac{1}{2},0,\frac{1}{2}$.
2	b	mmm	$0,0,\frac{1}{2}$; $\frac{1}{2},\frac{1}{2},0$.
2	a	mmm	$0,0,0$; $\frac{1}{2},\frac{1}{2},\frac{1}{2}$.

Fig. 1.42. Diagram to show the general equivalent positions and symmetry elements in the tetragonal space group $P4_2/m\,nm$. A new symbol here is that for the 4_2 axis, shown passing through the points $\frac{1}{2}$, 0 and 0, $\frac{1}{2}$. The general position that is marked by a split circle, say, open half-circle + and 'comma' half-circle − , indicate two points related by the m-plane parallel to the plane of the diagram. The coordinates of the more symmetrical special equivalent positions are listed. Note also the diagonal glide-planes interleaving the diagonal m-planes

Tetragonal space groups

Although tetragonal space groups present further complexities, they introduce new features some of which we should consider. We shall use the space group $P4_2/m\ nm$ as an example. Figure 1.42 illustrates its general equivalent positions and the symmetry elements. This space group is related to point group $4/m\ mm$ and, again, Table 1.2 reveals the expected orientation of the symmetry elements. Thus, the m-plane in the plane of the diagram is normal to the 4_2 axis (z), an n-glide plane is normal to x (and to its symmetry-related direction of y), and m-planes are normal to [110] and [1$\bar{1}$0]. Unlike point group $4/m\ mm$, the symmetry elements do not all pass through a single point, because they are offset by the translational symmetry present in the space group. A selection of the special equivalent positions is listed for this space group, and we shall meet it again in the rutile-type structures, which are relatively common for structures of the general formula MX_2.

Among the new features encountered with tetragonal space groups are the equivalence of the x and y directions, and the diagonal symmetry planes normal to the [110] and [$\bar{1}$10] directions. We note that all sets of symmetry planes and axes indicated by a space-group symbol are interleaved, or 'halved', by other similar symmetry elements in the expression of the space group.

Hexagonal space groups

In the hexagonal system, we consider the space group $P6_3mc$ which, by addition of a centre of symmetry becomes $P6_3/m\ mc$, a space group that represents the close-packed hexagonal structure adopted by many metallic elements in a close-packed, twelve-fold coordination.

Figure 1.43 shows the general equivalent positions and symmetry elements for space group $P6_3mc$; the coordinates of the general and special equivalent positions are listed. The symmetry at the space group origin, on the 6_3 axis (z), is $3m$. As with previous examples, we can deduce the point groups of this space group as $6mm$, so that we obtain the relative orientations of the symmetry elements by comparison with Table 1.2.

We see that, in this space group, the symmetry elements 6_3, m and c pass through the origin, because the translation of the c-glide-plane is matched by that of the 6_3 axis. The c-glide-planes are interleaved by n-glide planes, whereas the m-planes are interleaved by glide-planes, indicated by dashed lines, that behave rather like a (or b)-glide lanes, but having the translations of the form $(p+q)/2$, where p and q are any two of the three equivalent vectors in the x,y plane of which a and b are two..

EXAMPLE 1.3. Consider a hexagonal unit cell, as defined in Table 1.7, and 'centre' the a,b plane. (a) What system is now represented by this unit cell? (b) What is the symmetry at each point? (c) Choose a new unit cell and give its type, and the parameters in terms of a_H, b_H and c_H. The answers are at the end of the chapter.

Cubic space groups

Many simple substances crystallize in the cubic system with highly symmetrical space groups, such as $Fm3m$ and $F\bar{4}3m$. Drawings of the cubic space groups, especially those of higher symmetry, are very complex[38], because of the three-fold symmetry axes that are inclined to the plane of the diagram. Consequently, we shall be content to list a selection of the sets of special equivalent positions. Many of the cubic substances known have their component atoms in special positions of space groups in the cubic system. Table 1.8 lists these positions for space group $Fm3m$ and Table 1.9 those for $F\bar{4}3m$, both of which are of common occurrence.

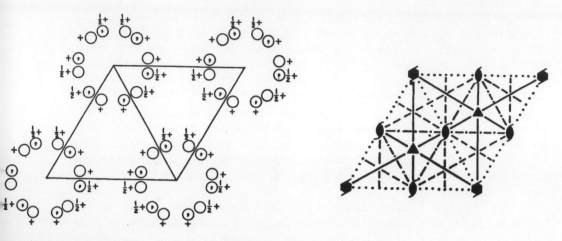

Origin on 6_3 ($3m$)

Number of positions	Wyckoff notation	Point symmetry	Coordinates of equivalent positions		
12	d	1	x, y, z;	$\bar{y}, x-y, z$;	$y-x, \bar{x}, z$;
			$\bar{y}, \bar{x}, z$;	$x, x-y, z$;	$y-x, y, z$;
			$\bar{x}, \bar{y}, \frac{1}{2}+z$;	$y, y-x, \frac{1}{2}+z$;	$x-y, x, \frac{1}{2}+z$;
			$y, x, \frac{1}{2}+z$;	$\bar{x}, y-x, \frac{1}{2}+z$;	$x-y, \bar{y}, \frac{1}{2}+z$.
6	c	m	$x, \bar{x}, z$;	$x, 2x, z$;	$2\bar{x}, \bar{x}, z$;
			$\bar{x}, x, \frac{1}{2}+z$;	$\bar{x}, 2\bar{x}, \frac{1}{2}+z$;	$2x, x, \frac{1}{2}+z$.
2	b	$3m$	$1/3, 2/3, z$;	$2/3, 1/3, 1/2+z$.	
2	a	$3m$	$0, 0, z$;	$0, 0, 1/2+z$.	

Fig. 1.43. Diagrams to show the general equivalent positions and symmetry elements in space group $P6_3mc$. The coordinates of the complete general and special equivalent positions are listed in their sets. The dashed lines represent vertical glide-planes, with translations of the form $(a + b)/2$, that interleave the m-planes. Note also the symbol for the 6_2 axis.

Sodium chloride and most of the other alkali-metal halides, and the chalcides (oxides, sulfides, selenides and tellurides) of the alkaline-earth metals are some of the compounds that crystallize with space group $Fm3m$.

Representatives of space group $F\bar{4}3m$ include zinc blende, the cubic form of zinc sulfide, the iodides of silver and copper(I), and a number of binary compounds, such as BN, InSb and SiC, that we shall discuss later on.

We have completed now our introductory survey of the symmetry of molecules and crystals. We shall draw upon the above data in the work of later chapters. Reference may be made as necessary to the definitive work[4] on space groups in studying structures, and further general information of this topic is readily available[2-6] for detailed study. It may be found helpful to refer back frequently to Table 1.2 and Figure 1.6, because the *orientations* of symmetry elements in space groups parallel those in their corresponding point groups, although the symmetry elements in the space groups may be displaced from their point-group *positions* by the presence of translational symmetry.

Table 1.8 Selected special equivalent positions for space group *Fm3m*

Origin at centre (*m3m*)

Number of positions	Wyckoff notation	Point symmetry	Coordinates of equivalent positions
			$(0,0,0\,;\ 0,\frac{1}{2},\frac{1}{2}\,;\ \ \frac{1}{2},0,\frac{1}{2}\,;\ \frac{1}{2},\frac{1}{2},0)+$
32	*f*	3*m*	$x,\ x,\ x\,;\ x,\bar{x},\bar{x}\,;\ \ \bar{x},x,\bar{x}\,;\ \bar{x},\bar{x},x\,;$ $\bar{x},\bar{x},\bar{x}\,;\ \bar{x},x,x\,;\ \ x,\bar{x},x\,;\ x,x,\bar{x}\,.$
24	*e*	4*mm*	$x,0,0\,;\ 0,x,0\,;\ 0,0,x\,;$ $\bar{x},0,0\,;\ 0,\bar{x},0\,;\ 0,0,\bar{x}\,.$
24	*d*	*mmm*	$0,\frac{1}{4},\frac{1}{4}\,;\ \frac{1}{4},0,\frac{1}{4}\,;\ \frac{1}{4},\frac{1}{4},0\,;$ $0,\frac{1}{4},\frac{3}{4}\,;\ \frac{3}{4},0,\frac{1}{4}\,;\ \frac{1}{4},\frac{3}{4},0\,.$
8	*c*	$\bar{4}\,3m$	$\frac{1}{4},\frac{1}{4},\frac{1}{4}\,;\ \frac{3}{4},\frac{3}{4},\frac{3}{4}\,.$
4	*b*	*m3m*	$\frac{1}{2},\frac{1}{2},\frac{1}{2}\,.$
4	*a*	*m3m*	$0,0,0\,.$

Table 1.9 Selected special equivalent positions for space group $F\bar{4}\,3m$

Origin at $\bar{4}\,3m$

Number of positions	Wyckoff notation	Point symmetry	Coordinates of equivalent positions
			$(0,0,0\,;\ 0,\frac{1}{2},\frac{1}{2}\,;\ \ \frac{1}{2},0,\frac{1}{2}\,;\ \frac{1}{2},\frac{1}{2},0)+$
24	*g*	*mm2*	$x,\frac{1}{4},\frac{1}{4}\,;\ \frac{1}{4},x,\frac{1}{4}\,;\ \frac{1}{4},\frac{1}{4},x\,;$ $\bar{x},\frac{1}{4},\frac{3}{4}\,;\ \frac{3}{4},\bar{x},\frac{1}{4}\,;\ \frac{1}{4},\frac{3}{4},\bar{x}\,.$
24	*f*	*mm2*	$x,0,0\,;\ 0,x,0\,;\ 0,0,x\,;$ $\bar{x},0,0\,;\ 0,\bar{x},0\,;\ 0,0,\bar{x}\,.$
16	*e*	3*m*	$x,x,x\,;\ x,\bar{x},\bar{x}\,;\ \bar{x},x,\bar{x}\,;\ \bar{x},\bar{x},x\,.$
4	*d*	$\bar{4}\,3m$	$\frac{3}{4},\frac{3}{4},\frac{3}{4}\,.$
4	*c*	$\bar{4}\,3m$	$\frac{1}{4},\frac{1}{4},\frac{1}{4}\,.$
4	*b*	$\bar{4}\,3m$	$\frac{1}{2},\frac{1}{2},\frac{1}{2}\,.$
4	*a*	$\bar{4}\,3m$	$0,0,0\,.$

ANSWERS TO EXAMPLES 1

1.1. (a) <u>Cube</u>. Three four-fold rotation axes (4) along x, y and z, or normal to the faces of the cube; four-fold roto-inversions axes ($\bar{4}$) and two-fold rotation axes (2) coincide with the four-fold axes. Four three-fold rotation axes (3) at $\cos^{-1}(1/\sqrt{3})$ to x, y and z, or along the body-diagonals of the cube; three-fold roto-inversion axes ($\bar{3}$) coincide with the three-fold axes. Six two-fold rotation axes (2) at 45° to x, y and z, or through the mid-points of opposite edges of the cube. Three mirror planes (m) normal to x, y and z, or parallel to the faces of the cube; they are equivalent to two-fold roto-inversion axes ($\bar{2}$) along x, y and z. Six mirror planes (m) normal to the two-fold axes, or through the face-diagonals of the cube. A centre of symmetry ($\bar{1}$) is present, at the geometrical centre of the cube.

(b) <u>Tetrahedron</u>. Three four-fold roto-inversion axes along x, y and z, or through the mid-points of opposite edges of the tetrahedron; two-fold rotation axes are coincident. Four three-fold rotation axes at $\cos^{-1}(1/\sqrt{3})$ to x, y and z, or through each apex. Six m-planes through one edge and the mid-point of the opposite edges of the tetrahedron; equivalent two-fold roto-inversion axes lie normally to each mirror plane.

1.2. (a) Cube, $m3m$. (b) Tetrahedron, $\bar{4}3m$.

1.3. (a) The symmetry is degraded by the 'centring', and the system becomes orthorhombic. (b) The symmetry at each lattice point is now mmm. (c) $a_O = a_H/2 - b_H/2$; $b_O = a_H/2 + b_H/2$. Thus, $a_O = \frac{1}{2}a_H \sqrt{3}$, and $b_O = a_H/2$. The 90° angle is confirmed, because $(a_H/2 - b_H/2)\cdot(a_H/2 + b_H/2) = 0$.

PROBLEMS 1

1.1. How many different point groups can be obtained from the combination 422 by replacing one or more of the rotation symmetry elements by the corresponding roto-inversion symmetry element?

1.2. Determine the point groups for the chemical species shown in Figures 1.2 to 1.4. If you are using the program *SYMH*, allocate the model numbers 84, 16 and 56, respectively, to them.

1.3. Which of the following descriptions represent Bravais lattices?
(i) Orthorhombic B;
(ii) Tetragonal A;
(iii) Triclinic I.

1.4. A monoclinic F lattice unit cell has the dimensions $a = 0.600$ nm, $b = 0.700$ nm, $c = 0.800$ nm, $\beta = 110.0°$. Show that an equivalent monoclinic C unit cell, with an *obtuse* β-angle, can represent the same lattice, and calculate its dimensions. What is the ratio of the volume of the C cell to that of the F cell?

1.5. Calculate the length of $[31\bar{2}]$ in both unit cells of Problem 1.4.

1.6. The relationships $a \neq b \neq c$, $\alpha \neq \beta \neq 90°$, 120°, $\gamma = 90°$ may be said to represent a 'diclinic' system. Is it a new system? If so, how so, and if not, why not?

1.7. Carry out the following exercise with a tetragonal P lattice unit cell:
(i) Centre the B faces. Comment on the result.
(ii) Centre the A and B faces. Comment on the result.
(iii) Centre all the faces. What conclusions can you draw now?

1.8. For each of the space groups $P2/c$, $Pca2_1$, $Cmcm$ and $P\bar{4}2_1c$:
(i) Write the crystal system and point group of the space group;
(ii) List the *full* meaning conveyed by the symbol.

1.9. Consider the following questions relating to space group $P2_1/c$.
(i) To what crystal system does it belong?
(ii) What are the orientations of the symmetry elements with respect to the x, y and z axes?

1.10. The monoclinic structure unit cell of gypsum, $CaSO_4·2H_2O$, has been selected in different ways:

	a/nm	b/nm	c/nm	β/deg
(i)	1.051	1.515	0.655	151.72
(ii)	0.567	1.515	0.655	118.58
(iii)	1.051	1.515	0.628	99.30

(a) Draw a projection of the Bravais lattice, using either data (i) or (ii), on to the x,z plane; a scale of 10 mm = 0.1 nm is suitable. It is convenient to use a transparent guide-sheet, and to draw a grid of lines to locate the lattice points. Then prick through the points on to another sheet of paper and, on it, outline the three unit cells. What are the ratios of the volume of cell (iii) to those of cells (I) and (ii)? Which unit cell would you choose? Give reasons.

1.11. Draw a cubic F unit cell and outline a rhombohedron within it, making the unique axis of the rhombohedron coincide with $[\bar{1}11]$ in the cube. What the unit-cell dimensions of the rhombohedron in terms of the cube unit cell side length a_C?

1.12. (a) Draw diagrams to show the general equivalent positions and symmetry elements in space group $Pma2$. (b) List the coordinates of the general and special equivalent positions in their sets. (c) How might this space group accommodate four or two molecules of 1,2-difluorobenzene?

1.13. By drawing, or otherwise, show the equivalence of tetragonal F and tetragonal I lattices.

CHECKLIST 1
At the end of this chapter, you should be able to:
1. View stereoscopic pairs of drawings correctly;
2. Understand symmetry operations and symmetry elements;
3. Recognize symmetry elements in finite bodies, specifically molecules and crystals;
4. Know the conventional reference axes for crystals;
5. Understand point groups and crystal systems;
6. Use the Hermann-Mauguin symmetry notation;
7. Determine the point group of a model of a molecule or crystal;
8. Understand lattices, unit cells and centring;
9. Use the notation for unit cells (lattices) and understand fractional coordinates;
10. Show the equivalence, or otherwise, of two given unit cells, e,g, monoclinic P and B;
11. Appreciate the sharing of lattice points among adjacent unit cells (P or centred);
12. Transform unit cell vectors (without change of origin);
13. Calculate transformed unit-cell dimensions by simple vector methods, or otherwise;
14. Understand directions and calculate their magnitudes;
15. Show how crystal rotational symmetry is limited to the degrees 1, 2, 3, 4 and 6;
16. Understand space group, general and special equivalent positions, Wyckoff's notation for sets of equivalent positions; screw axis, glide plane;
17. Construct a space group from a lattice and a motif of point-group symmetry;
18. Be familiar with example of space groups in different crystal systems, and use space-group tables.

2

Looking at ionic structures

2.1 INTRODUCTION

We have seen in the previous chapter that the basis of any crystal structure is a Bravais lattice. When a given chemical entity is arranged in an identical orientation at or around each Bravais lattice point, a structure is built up that may be described by a space group. For example, we can describe the structure of sodium chloride, Figure 2.1, in two ways.

Sodium chloride has a face-centred cubic structure unit cell of side 0.564 nm and crystallizes in space group $Fm3m$. Its density is 2165 kg m^{-3}; hence, we find the number of formula-entities in the unit cell to be $2165 \times (0.564 \times 10^{-9})^3 / [(22.99 + 35.45) \times 1.6605 \times 10^{-27}]$, which evaluates to 4.003, or 4 to the nearest integer. From Table 1.8, we see that the Na$^+$ and Cl$^-$ ions must occupy the two sets of special positions labelled a and b. Thus, if Cl$^-$ is in a, Na$^+$ is in b, or *vice versa*.

Alternatively, if we associate pairs of sodium and chloride ions in the correct and identical orientation at each point of the F cubic lattice, Figure 1.26, we obtain again the structure of sodium chloride. It is also sometimes spoken of as two interpenetrating cubic F lattices, one of Na$^+$ ions and the other of Cl$^-$ ions, but this terminology is not recommended: each crystal structure is based on a single lattice.

Although a space group is an infinite set of symmetry elements, we can apply the theory with success to finite structures because, in any crystal under examination, the number of unit cells is very large. For example, if we consider a cube of sodium chloride of side, say, 0.05 mm, then the number of unit cells in that cube is $(0.05 \times 10^{-3})^3/(0.564 \times 10^{-9})^3$, or approximately 7×10^{14}.

There is no unique way in which the multitude of known structures may be subdivided, but a method of classification is desirable and useful. We shall adopt a procedure based on the principal type of interparticle force responsible for cohesion of the crystal structural units in the *solid* state. Thus, we obtain four classes that may be termed ionic structures, covalent structures, metallic structures and molecular, or van der Waals, structures. Even so, we shall find that certain compounds do not fall clearly into one or other of these classes, so that the final choice is, to some extent, subjective. In this chapter we consider ionic structures.

2.2 IONIC STRUCTURES

The structures that we term ionic are formed generally between two species of widely different **electronegativity**. Electronegativity refers to the tendency of an atom to attract electrons in compound formation; it may be quantified on a relative scale[7]; Table 2.1 lists the Pauling[8] electronegativities for some elements (see also Section 3.3.1).

In forming an ionic compound, one species, typically a metal, forms a positive ion by the loss of one electron or more, and the other species, typically a non-metal acquires one electron or more and becomes negatively charged. Then, a Coulombic attractive energy exists between them that is inversely proportional to the distance between the ions and directly proportional to their charges. Specifically, if two ions of numerical charges q_+ and q_- (including their signs) are distant r apart, the Coulombic, electrostatic energy of

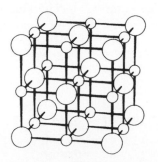 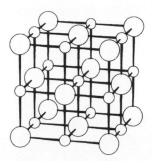

Fig. 2.1. Stereoview of the unit cell and environs of the sodium chloride structure type; the circles in order of decreasing size represent Cl^- and Na^+. The space group is $Fm3m$, with Cl^- in Wyckoff a and Na^+ in b positions. The positions of the sodium and chlorine species can be interchanged; it would be equivalent to a change of origin.

Table 2.1 Electronegativities for some elements

H						
2.20						
Li	Be	B	C	N	O	Fl
0.98	1.57	2.04	2.55	3.04	3.44	3.98
Na	Mg	Al	Si	P	S	Cl
0.93	1.31	1.61	1.90	2.19	2.58	3.16
K	Ca	Ga	Ge	As	Se	Br
0.82	1.00	1.81	2.01	2.18	2.55	2.96
Rb	Sr	In	Sn	Sb	Te	I
0.82	0.95	1.78	1.96	2.05	2.36	2.66
Cs	Ba	Tl	Pb	Bi		
0.79	0.89	2.04	2.33	2.02		
Zn	Cd	Hg				
1.65	1.69	2.00				

attraction u_E between them is given by

$$(2.1) \qquad u_E = q_+ q_- e^2/(4\pi\varepsilon_0 r) \; .$$

where e is the electronic charge and ε_0 is the permittivity of a vacuum. In an extended array of ions, as in a crystal, it is necessary to take account of the effect of all ions in the structure in obtaining the true electrostatic energy. We write the cohesive energy of the structure, often called the **lattice energy**, $U_E(r)$ as

$$(2.2) \qquad U_E(r) = A q_+ q_- e^2/(4\pi\varepsilon_0 r) \; .$$

where A is a constant for the structure type known as the **Madelung** constant.

Table 2.2 Selected Madelung constants

Structure type	q_+	q_-	A	$-q_+q_-A$	$-rU_E/\text{kJ mol}^{-1}$
CsCl	1	−1	1.7627	1.7627	244.9
NaCl	1	−1	1.7476	1.7476	242.8
α-ZnS (Wurtzite)	2	−2	1.6407	6.5628	911.8
β-ZnS (Blende)	2	−2	1.6381	6.5523	910.4
CaF_2	2	−1	2.5194	5.0388	700.1
TiO_2 (Rutile)	4	−2	2.3851	19.0791	2650.7
BeF_2	2	−1	2.2011	4.4022	611.6
β-SiO_2	4	−2	2.2011	17.6088	2446.4
CdI_2	2	−1	2.1921	4.3842	608.9

2.2.1 Madelung constant

Consider the sodium chloride structure type, Figure 2.1, and take the central Na^+ ion as a reference origin. The six nearest Cl^- neighbours give rise to an attractive energy of $-6e^2/(4\pi\varepsilon_0 r)$, since $q_+q_- = -1$, where r is the shortest $Na^+\cdots Cl^-$ distance. The next nearest neighbours set up a Coulombic repulsion energy of $12e^2/(4\pi\varepsilon_0 r\sqrt{2})$, and the next nearest, at the corners of the cell, give rise to an attractive energy term $-8e^2/\{4\pi\varepsilon_0 r\sqrt{3}\}$, and so on. These terms form the series

(2.3) $$A = - [6 - 12/\sqrt{2} + 8/\sqrt{3} - \cdots] ,$$

which has a very slow rate of convergence, and is not suitable for calculation of the Madelung constant

In the approximation due to Evjen[40], a small portion of the structure is treated as a neutral block: potential energy falls off more rapidly with distance for a neutral block of structure than with a group carrying excess charge. Thus, in a single unit cell, each ion is weighted according to its share in the given unit cell, and we obtain the modified series

(2.4) $$A = - [(6 \times 1/2) - (12 \times 1/4)/\sqrt{2} + (8 \times 1/8)/\sqrt{3}] ,$$

which sums to 1.46. For a unit cell of twice the size, the Evjen series sums to 1.75. The accurate value for the Madelung constant for this structure is $1.74756 \cdots$. A selection of Madelung constants is listed in Table 2.2, and Appendix 2 discusses their calculation.

In Table 2.2, the values of A refer to the structure type with unit charges at the ion sites. There is no unique value of r for the rutile structure, for example, and the smallest interionic distance in the structure (0.1945 nm) is used in calculating the Madelung constant. A final column gives the values for the electrostatic energy when divided by the appropriate value of r in nm; thus, for NaCl, $U_E(r) = -244.9/0.282 = -868$ kJ mol^{-1}.

2.2.2 Lattice energy: point-charge model

In all solids, it follows from Earnshaw's theorem in electrostatics that the energy of attraction between species is balanced by an energy of repulsion in achieving an equilibrium condition. A suitable repulsion potential energy takes the form $B \exp(-r/\rho)$, where B is a constant for the structure type and ρ is a constant for the substance itself, and is related to its compressibility. Thus, we may write for the lattice energy $U(r)$,

(2.5) $U(r) = A'/r + B \exp(ar)$,

where $A' = q_+q_-e^2/(4\pi\varepsilon_0)$ and $a = -1/\rho$. We identify an energy minimum at the equilibrium distance r_e, and from (2.5) we obtain

(2.6) $dU/dr = -A'/r^2 + aB \exp(ar)$.

At $r = r_e$, $dU/dr = 0$ and we eliminate B as

(2.7) $B = A'/[r_e^2 \, a \exp(ar_e)]$.

Using the expression for B from (2.7) in (2.5), and inserting the components of A' and a, and multiplying by $10^{-3}L$ so that the result shall be in the usual units of kJ mol^{-1},

(2.8) $U(r_e)/\text{kJ mol}^{-1} = 10^{-3}LA \, q_+q_-e^2(1- \rho/ r_e)/(4\pi\varepsilon_0 r_e)$,

where L is the Avogadro constant and q_+ and q_- include their signs. In many ionic compounds ρ/r_e is sufficiently close to 0.1 for (2.8) to be a useful equation for obtaining values of their lattice energies. Inserting the constants into (2.8) and assuming 0.1 for the value of ρ/r_e, with r_e in nm, we obtain the approximate lattice energy equation for an ionic compound as

(2.9) $U(r_e)/\text{kJ mol}^{-1} = 125Aq_+q_-/(r_e/\text{nm})$.

For sodium chloride, $A = 1.7476$, $q_+q_- = -1$ and $r_e = 0.282$ nm, so that $U(r_e) = -775$ kJ mol^{-1}. Refinement of the point-charge model, to include induced-dipole–induced-dipole and induced-dipole–induced-quadrupole terms[7] leads to $U(r_e) = -775$ kJ mol^{-1}. The good agreement with the result from the approximate (2.9) is fortuitous, and one should not expect this precision from (2.9) generally.

2.2.3 Lattice energy: experimental approach

An alternative, independent method for calculating lattice energies, at least for the alkali-metal halides and certain other compounds, makes use of thermodynamic parameters that have been determined by experiment. If we set out to make sodium chloride, for example, we introduce metallic sodium into an atmosphere of gaseous chlorine and irradiate it so as to initiate the reaction, whereupon a white solid, sodium chloride, is formed. We express the process (enthalpy ΔH_f°) by the equation

(2.10) $Na(cr) + \tfrac{1}{2}Cl_2(g) \longrightarrow Na^+Cl^-(cr)$.

This process is accompanied by the liberation of the enthalpy of formation of the crystalline solid, but it tells us nothing about the nature of the interatomic forces present.

   We can represent the formation of a compound MX by a number of stages, as shown in Figure 2.2. The metal $M(cr)$ in its standard state (298.15 K and 101.325kPa) is converted to the gaseous ion at the same temperature and presssure, through sublimation (enthalpy ΔH_s°) and ionization (enthalpy ΔH_i°). Similarly, the gaseous non-metal $X_2(g)$ is converted to a gaseous anion through dissociation (enthalpy ΔH_d°) and electron affination (enthalpy ΔH_e°). These four processes involve supplying enthalpy to the system, a total of 379.1 kJmol^{-1} in the case of sodium and chlorine, which is thermo-

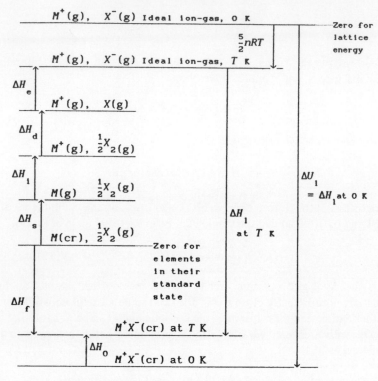

Fig. 2.2. Thermodynamic (Born-Haber) cycle for lattice energy and lattice enthalpy.

dynamically unfavourable. However, the reaction is driven forward by release of the lattice enthalpy $\Delta H_1{}^{\circ}$.

The lattice *energy* is defined thermodynamically as the change in internal energy ΔU_1 (or $\Delta U_1{}^{\circ}$ under standard state conditions) for the process

$$(2.11) \qquad\qquad Na^+(g) + Cl^-(g) \longrightarrow Na^+Cl^-(cr)$$

at 0 K; the reference zero for lattice energy is the ideal ion-gas at 0 K, so that $\Delta U_1{}^{\circ}(0) = \Delta H_1{}^{\circ}(0) = U_1{}^{\circ}(0)$. At any temperature T, $\Delta U_1{}^{\circ}(T) = \Delta H_1{}^{\circ}(T) - p[V(cr) - V(g)] = \Delta H_1{}^{\circ}(T) + nRT$, because $V(g) \gg V(cr)$; n is the total number of moles per mole of ion-gas, two in the case of a compound MX, and. R is the universal gas constant. From the cycle, the lattice enthalpy at a temperature T is given by

$$(2.12) \qquad\qquad \Delta H_1{}^{\circ} = \Delta H_f{}^{\circ} - \Delta H_s{}^{\circ} - \Delta H_i{}^{\circ} - \tfrac{1}{2}\Delta H_d{}^{\circ} - \Delta H_e{}^{\circ} + 5nRT/2,$$

The term $5nRT/2$ is the enthalpy change $\int_0^T C_{P(g)}\,dT$, which is $5RT/2$ per mole for an ideal gas. The quantity $\Delta H_0{}^{\circ}$ is $\int_0^T C_{P(cr)}\,dT$, which is evaluated from heat capacity data on the crystal MX. Thus, we obtain the lattice energy at T as

$$(2.13) \qquad\qquad \Delta U_1{}^{\circ} = \Delta H_f{}^{\circ} - \Delta H_s{}^{\circ} - \Delta H_i{}^{\circ} - \tfrac{1}{2}\Delta H_d{}^{\circ} - \Delta H_e{}^{\circ} + 7nRT/2.$$

Inserting the data appropriate to sodium chloride, we obtain $\Delta U_l^e = -774 \pm 2$ kJ mol^{-1}, which provides excellent agreement with electrostatic model, and serves to confirm its suitability as a method for ionic crystals. The useful interplay of the two approaches to lattice energy values in deriving parameters that cannot be measured experimentally, such as the electron affinity $O(g) + 2e^- \rightarrow O^{2-}(g)$, or the enthalpy term that combines dissociation and electron affination, such as $\Delta H_f(NO_3^-, g)$, in discussing compound stability, and in determining the charge distribution on complex ions, have been discussed extensively elsewhere[7,9,10,30].

2.3 CRYSTAL CHEMISTRY

A systematic approach to studying the relationship between the properties of solids and their internal structure was made possible first through the X-ray analysis of simple ionic compounds. Table 2.3 lists the equilibrium interionic distances in the alkali-metal halides, mostly of the sodium chloride structure type, from X-ray analysis.

Table 2.3 Equilibrium interionic distances r_e/nm for the alkali-metal halides

	Li	Δ	Na	Δ	K	Δ	Rb	Δ	Cs	Average, $\overline{\Delta}$
F	0.201	*0.030*	0.231	*0.036*	0.267	*0.015*	0.282	*0.018*	0.300	
Δ	*0.056*		*0.049*		*0.047*		*0.046*		*0.056*	***0.051***
Cl	0.257	*0.025*	0.282	*0.032*	0.314	*0.014*	0.328	*0.028*	0.356	
Δ	*0.018*		*0.017*		*0.015*		*0.015*		*0.016*	***0.016***
Br	0.275	*0.023*	0.298	*0.031*	0.329	*0.014*	0.343	*0.028*	0.371	
Δ	*0.025*		*0.025*		*0.024*		*0.023*		*0.024*	***0.024***
I	0.300	*0.023*	0.323	*0.030*	0.353	*0.013*	0.366	*0.029*	0.395	
Average, $\overline{\Delta}$	*0.025*		*0.033*		*0.014*		*0.026*[a]			

[a]It may be noted that CsCl, CsBr and CsI have the cesium chloride structure type at ambient conditions.

The values of Δ show that the differences between r_e for two halides of a given cation are almost independent of the nature of the cation, and *mutatis mutandis* for a given anion. These results can be explained by a model in which ions are spherical, each of a characteristic radius, such that the radii sums correspond to the equilibrium interionic distances. Thus, for potassium and sodium chlorides, for example, it is postulated that

$$r_e(KCl) - r_e(NaCl) = [r(K^+) + r(Cl^-)] - [r(Na^+) + r(Cl^-)] = r(K^+) - r(Na^+),$$

which is then independent of the halogen considered.

2.3.1 Ionic radii

The extraction of the individual ionic radii from the experimental parameter r_e has attracted attention since it was first carried out by Landé in 1920, because there is no unique method for making the division. Landé considered the six compounds listed in

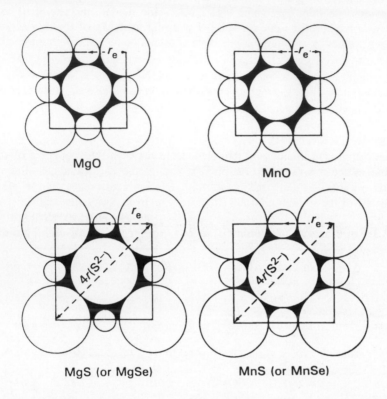

Fig. 2.3. Close-packed arrays of ions, as seen in projection on to a face of the cubic unit cell. Close contact exists across face diagonals for the sulfides and selenides, but not for the oxides.

Table 2.4 Interionic distance in compounds of the NaCl structure type

	r_e/nm		r_e/nm
MgO	0.210	MnO	0.222
MgS	0.260	MnS	0.261
MgSe	0.273	MnSe	0.273

Table 2.4, all of which have the sodium chloride structure type. The constancy of r_e for the sulfides and selenides was taken to indicate that the anions were in close contact, with the smaller cations occupying the interstices, as shown in Figure 2.3.

In the close-packed array of selenide anions, we have $2\,r_e\sqrt2 = 4r(Se^{2-})$, so that $r(Se^{2-})$ = 0.193 nm; similarly, $r(S^{2-})$ = 0.184 nm. The two oxides cannot be treated in this way, because their interionic distances indicate an uncertainty about the close-packing in these compounds. However, if we accept the additivity of ionic radii, then other radii can be deduced from $r(Se^{2-})$, $r(S^{2-})$ and others; the Goldschmidt radii (1926) were deduced in this manner. Pauling (1930) showed that the radius of an ion is governed by the configuration of the outermost electrons in the corresponding atom. For an isoelectronic series of ions, he gave the relationship

(2.14) $r_i = c/(Z_i - \sigma)$, where c is a constant for an isoelectronic series of ions, Z_i is the atomic number of the ith ion of radius r_i, and σ is Slater's screening constant[7] which allows for the screening effect of the outer electrons on the nuclear charge of the ion. Considering NaF, for example, we have from (2.14)

$$r(Na^+) = c/(11 - 4.15)$$

and

$$r(F^-) = c/(11 - 4.15),$$

whence

$$r(Na^+)/r(F^-) = 0.708.$$

Since $r_e(NaF) = 0.231$ nm and $r_e = r_+ + r_-$, $r(Na^+) = 0.096$ nm, and $r(F^-) = 0.135$ nm.

By applying Landé's method to lithium iodide, for which close packing of anions seemed certain, Ladd obtained[13] a set of radii very similar to those deduced from direct measurements on X-ray electron density maps of alkali-metal halides[14]. These electron density maps were sufficiently detailed for the zero-density contour to be identified around each ion, so that a direct measurement of an ionic radius was feasibile. Table 2.5 lists Pauling's and Ladd's ionic radii together with values from independent observations by Shannon and Prewitt[15]. The close correspondence between the Ladd and Shannon-Prewitt values tends to accord them the greater reliability. Except for the Be^{2+} and Mg^{2+}

Table 2.5 Radii/nm of some ionic species

	Ladd	Pauling	Shannon-Prewitt
Li^+	0.086	0.060	0.088
Na^+	0.112	0.095	0.116
K^+	0.144	0.133	0.152
Rb^+	0.158	0.148	0.163
Cs^+	0.184	0.169	0.184
NH_4^+	0.166	0.148	-
Ag^+	0.127	0.126	0.129
Tl^+	0.154	0.140	-
Be^{2+}	0.048	0.031	0.041
Mg^{2+}	0.087	0.065	0.086
Ca^{2+}	0.118	0.099	0.114
Sr^{2+}	0.132	0.113	0.130
Ba^{2+}	0.149	0.135	0.150
H^-	0.139	0.208	-
F^-	0.119	0.136	-
Cl^-	0.170	0.181	-
Br^-	0.187	0.195	-
I^-	0.212	0.216	-
O^{2-}	0.125	0.140	-
S^{2-}	0.170	0.184	-
Se^{2-}	0.181	0.198	-
Te^{2-}	0.197	0.221	-

Table 2.6 Radii adjustments due to coordination

Coordination number	Adjustment
4	Standard – 5%
6	Standard
8	Standard + 3%

cations (4-coordination), the radii in Table 2.5 applies to 6-coordination. Measurements on polymorphic species show that the radii are dependent upon coordination number. Taking 6-coordination as the standard, the adjustments given in Table 2.6 apply.

2.3.2 Radius ratio and ionic *MX* structure types
The radius ratio R is defined by

(2.15) $$R = r_+/r_- ,$$

and may be used as a *guide* in predicting structure type. We shall investigate the radius ratio and, at the same time, study the ionic structure types of the general formula *MX*.

Consider the cesium chloride structure type, Figure 2.4, in which the coordination pattern is 8:8, that is, each species is surrounded by eight of the other kind. The structure has the space group *Pm3m*, and the ions occupy the Wyckoff sets *a* and *b*, Table 1.8 (without *F*-centring). We note that the structure is not body-centred, because that would require the species at the corners and the centre of the unit cell to be one and the same type. Two important classes of common compounds crystallize with the cesium chloride structure. In one class, we find the halides of the larger singly-charged cations, Cs^+, Tl^+ and NH_4^+, and in the other there are intermetallic compounds (binary alloys), such BeCu and CuZn. A change in external conditions that would lead to closer packing, that is, an increase in pressure or decrease in temperature, can bring about a transformation from the sodium chloride structure to that of cesium chloride. Thus, rubidium chloride at 83K or *ca* 7300 bar transforms from the sodium chloride to the cesium chloride structure type.

If the ions in the CsCl structure are of such relative sizes that the anions at the corners of the unit cell of side *a* are simultaneously in contact with one another and with the central, smaller cation, it follows that

$$2(r_+ + r_-) = a\sqrt{3} = 2 \, r_-\sqrt{3} ,$$

so that the radius ratio for maximum contact in eight-fold coordination R_8 becomes

(2.16) $$R_8 = 0.732 .$$

If, now, the cation be made smaller while maintaining the anionic radius constant, contact is lost between the cation and anions. There arises a separation of charged regions, even though the interionic distance is unaltered, and we may ask if a more stable structure can be obtained by a change in the pattern of coordination. It transpires that the sodium chloride 6:6 structure type, Figure 2.1 (see also Table 1.8), becomes more stable with a smaller value of the radius ratio. By reference to Figure 2.3, taking MnS as

$$2(r_+ + r_-) = a\sqrt{2} = 2 \, r_-\sqrt{2} ,$$

an example, so that

(2.17) $R_6 = 0.414$.

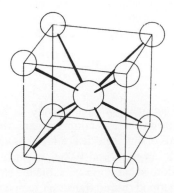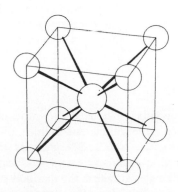

Fig. 2.4. Stereoview of the unit cell and environs of the CsCl structure type; circles, in order of decreasing size represent Cs^+ and Cl^-. The space group is $Pm3m$, with Cl^- in Wyckoff a and Cs^+ in b positions. We note that the structure is not *body-centred*, because the species at the centre of the unit cell is different from that at the corners.

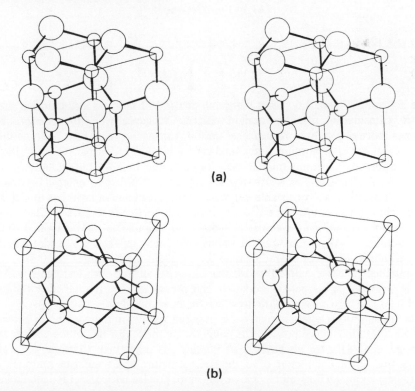

(a)

(b)

Fig. 2.5. Stereoviews of the unit cell and environs of the structures of zinc sulfide; circles in order of decreasing size represent Zn and S. (a) Wurtzite (α-ZnS): the space group is $P6_3mc$, with Zn^{2+} in Wyckoff a and S^{2-} in b. (b) Zinc blende (β-ZnS): the space group is $F\overline{4}3m$ with Zn^{2+} in Wyckoff a positions and S^{2-} in c. The $[ZnS_4]$, or $[Zn_4S]$, structural unit is present in each.

The largest class of MX compounds has the sodium chloride structure type: most alkali-metal halides, hydrosulfides and cyanides, silver halides; nearly all alkaline-earth metal chalcides; transition-metal monoxides; many binary compounds of nitrogen, phosphorus, arsenic, antimony and bismuth, with some trivalent metals, notably lanthanons; binary intermetallic compounds (see also Section 4.6.3).

In a similar manner, we can obtain the radius ratio for the 4:4 coordinated structures of zinc sulfide, Figure 2.5 (see also Table 1.9):

$$(2.18) \qquad\qquad R_4 = 0.225 \ .$$

Compounds that have the structure type of wurtzite or zinc blende include the chalcides of beryllium, and halides, oxides and nitrides of transition-type and the more electronegative metals. Energetically, the two forms of zinc sulfide are very similar; their difference in electrostatic energy is less that 0.2%. Both α- and β-ZnS occur in nature and, although wurtzite is the thermodynamically more stable form, zinc blende shows no tendency to transform to wurtzite with time; the thermodynamic transition temperature is ca 1300 K.

The change in coordination with radius ratio may be given a partial quantitative explanation by means of the lattice energy equation. We use (2.9), which we write as

$$U(r_e)/\text{kJ mol}^{-1}= \alpha/r_e ,$$

where $\alpha = 125Aq_+q_-$. Let $q_+ = 1$, $q_- = -1$ and $r_e = r_+ + r_-$. With r_- fixed, we have

$$(2.19) \qquad\qquad U(r_e)/\text{kJ mol}^{-1} = \beta/(R+1) ,$$

where $\beta = \alpha/r_-$. Figure 2.6 shows the graph of $U(r_e)/\text{kJ mol}^{-1}$ as a function of the radius ratio R. Considering first the cesium chloride structure type, the energy decreases (becomes more negative) as R decreases until it reaches the value of 0.732. At this value of R, the ions are in maximum contact and cannot become closer packed, even though the central cation may become smaller than its enclosing hole. Consequently, r_e remains constant as R decreases, as shown by Figure 2.6. The lattice energy becomes more negative, indicating a more stable structure, if the coordination switches to 6:6, and the sodium chloride structure type is preferred as R decreases to 0.414. Then, the curve again becomes parallel to the abscissa, and similar arguments can be applied to the zinc sulfide structures. We summarize the radius ratio limits in Table 2.7, and Table 2.8 lists the experimental values of the radius ratio for the alkali-metal halides.

A number of these halides, those marked with an asterisk, maintains the sodium chloride structure at normal temperatures and pressures even though their radius ratios exceed 0.732. There are several factors that we must consider.

The radius ratio is a geometrical concept based simply on the packing of spheres, and so has its limitations. Furthermore, the point-charge model for the lattice energy equation developed in (2.8) takes no account of interactions other than Coulombic. Table 2.2 suggests that, for a given value of r_e, the CsCl structure type should be more stable than that of NaCl by by ca 7 kJ mol^{-1}. However, more precise lattice energy calculations that take into account dipole and quadrupole terms in the lattice energy equation[7] confirm the preference for the sodium chloride structure type, as shown by Table 2.8. Although these induced multipole terms make small contributions to the lattice energy, they become important when the electrostatic energies of two species are close.

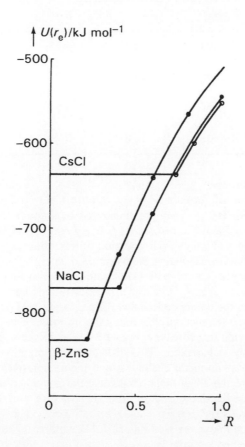

Fig. 2.6. Variation of lattice energy $U(r_e)$ with radius ratio R, at a constant value of r_-; the curves reflect the small difference in electrostatic energy between the cesium chloride and sodium structure types (see Table 2.2). The corresponding difference between α-zinc sulfide (not shown on this graph) and β-zinc sulfide is even smaller; both compounds contain the [ZnS_4], or [Zn_4S], structural unit.

Table 2.7 Radius ratio and *MX* structure type

R	Structure type
≥ 0.732	Cesium chloride
0.732–0.414	Sodium chloride
0.414–0.225	α/β-Zinc sulfide

There is also a certain covalent character associated with the bonding, even in these highly ionic structure types. In the sodium chloride structure type, the p orbitals (see Section 3.2.2ff) of adjacent ions are directed towards one another, thus facilitating overlap (see Section 3.2.5). In the structure type of cesium chloride this condition does not exist, because of the different pattern arising from the 8:8 coordination.

Table 2.8 Radius ratios in alkali-metal halides

	Li	Na	K	Rb	Cs
F	0.72	0.94[*]	0.83[*]	0.75[*]	0.65
Cl	0.51	0.66	0.85[*]	0.93[*]	0.92
Br	0.46	0.60	0.77[*]	0.84[*]	0.98
I	0.41	0.53	0.68	0.75[*]	0.87

2.3.3 Ionic MX_2 structure types

The common structure types of formula MX_2 are fluorite CaF_2, rutile TiO_2 and, less commonly, beryllium fluoride BeF_2. They are illustrated by Figures 2.7–2.9, and they show the coordinations 8:4, 6:3 and 4:2, respectively. They may be compared with the structures of cesium chloride, sodium chloride and zinc sulfide, Figures 2.4, 2.1 and 2.5, respectively, and they take the corresponding radius ratio limits.

Fluorite crystallizes in space group $Fm3m$ (see Table 1.8), with four formula-entities per unit cell. The calcium ions occupy Wyckoff a positions, and lie at the centre of eight fluoride ions that form the corners of a cube. The fluoride ions occupy Wyckoff c positions, and are coordinated tetrahedrally by four calcium ions. The $[CaF_8]$ structural units mirror the cesium chloride structure type, Figure 2.4, and it is this pattern that leads to the same limits here for the radius ratio. This structure is adopted by many halides of doubly-charged cations, the alkali-metal chalcides, and binary intermetallic compounds such as $AuAl_2$ and $IrSn_2$. In the alkali-metal chalcides, the positions of the anions and cations are reversed, compared with the calcium fluoride structure, and these compounds have been termed *anti-fluorite* structures.

Rutile is tetragonal, space $P4_2/m\ nm$ (see Figure 1.42), with two formula-entities per unit cell. The titanium species occupy Wyckoff a positions, with oxygen in f and determined by one parameter, x (0.3056). The titanium ions are coordinated octahedrally by six oxygen anions, and each oxide ion is linked to three titanium ions in an isosceles triangular arrangement. The coordination around the titanium parallels that of the sodium

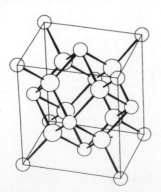
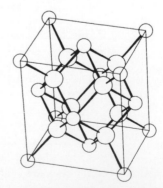

Fig. 2.7. Stereoview of the unit cell and environs of the fluorite structure; the circles in order of decreasing size represent F⁻ and Ca²⁺. The space group is $Fm3m$, with Ca²⁺ in Wyckoff a and F⁻ in c positions. The 8-coordination of F⁻ around Ca²⁺ mirrors the pattern of the CsCl structure type.

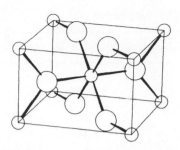 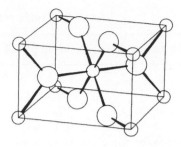

Fig. 2.8. Stereoview of the unit cell and environs of the rutile structure; the circles in order of decreasing size represent O^{2-} and Ti^{4+}. The space group is $P4_2/m\,nm$, with Ti^{4+} in Wyckoff a and O^{2-} in f positions.

chloride structure. The ionic species have been given formal charges, but the actual values are smaller, probably between two and three for titanium.

The rutile structure is adopted by many fluorides of transition-type metals and magnesium, and by dioxides of some transition-type metals and those of periodic groups 13 and 14 (see Section 3.2.3). Titanium dioxide exists in two other forms, anatase and brookite; the order of thermodynamic stability is

$$\text{Anatase} > \text{Rutile} > \text{Brookite}.$$

The structure of beryllium fluoride is similar to that of wurtzite and to that of the diamond allotrope of carbon. β-Cristobalite SiO_2 is sometimes included in this class: it has the same structure type but may be more justly placed in the covalent class, with other forms of silicon dioxide. Beryllium fluoride and β-cristobalite crystallize in the cubic system with space group $Fd3m$ and eight formula-entities per unit cell. In beryllium fluoride, the beryllium species are coordinated tetrahedrally by four fluoride ions, and each fluoride ion is linked non-linearly to two beryllium ions. The charges on the ions may be reduced slightly from the formal values, owing to a contribution from covalent bonding, more particularly so in the cristobalite structure.

2.3.4 Radius ratio and polarization in MX_2 structure types

Table 2.9 lists the radius ratios for a selection of MX_2 compounds. It is notable that these compounds obey the radius ratio limits without exception. Table 2.2 shows that the

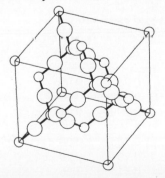 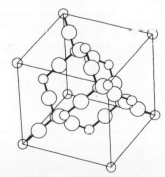

Fig. 2.9. Stereoview of the unit cell and environs of the beryllium fluoride structure; the circles in order of decreasing size represent F^- and Be^{2+}. The space group is $Fd3m$, with Be^{2+} in Wyckoff a and F^- in c positions.

Table 2.9 Radius ratios in MX_2 compounds

Fluorite (> 0.732)	Rutile (0.732–0.414)	Beryllium fluoride (0.414–0.225)	
BaF_2 1.25	$CaCl_2$ 0.69	BeF_2	0.23
SrF_2 1.11	$CaBr_2$ 0.63	$(SiO_2$	0.30)
$BaCl_2$ 0.88	MgF_2 0.73		
CaF_2 0.99	MnF_2 0.68		
$SrCl_2$ 0.78	ZnF_2 0.62		

differences in electrostatic energy among these structure types are considerably greater than those between the MX types. The relatively small contributions from non-ionic binding energy are insufficient to hold sway over the strong electrostatic component of energy in these MX_2 structures.

The non-ionic contributions to the lattice energy are often referred to as polarization effects. Polarization may be considered as a distortion of the electron density of an ion due to the field of its neighbours. The polarizing power p of an ion i may be measured by its electric field ($q_i e^2/4\pi\varepsilon_0 r_i$), so that it increases with both a decrease in radius r_i and an increase in ionic charge q_i; the polarizability α_i of a species i is proportional to its volume, so that it increases with an increase in ionic radius. Thus, we can obtain series of relative polarizing power p and polarizability α for a selection of species:

$$p: \quad Be^{2+} > Mg^{2+} > O^{2-} > Ba^{2+} > Li^+ > S^{2-} > Se^{2-} > Na^+ > Te^{2-} > F^- > K^+ > Rb^+ > Cs^+ > Cl^- > Br^- > I^-$$

$$\alpha: \quad Te^{2-} > Se^{2-} > I^- > S^{2-} > Br^- > Cs^+ > Cl^- > Ba^{2+} > O^{2-} > Rb^+ > K^+ > F^- > Na^+ > Mg^{2+} > Li^+ > Be^{2+}$$

The displacement of the centres of gravity of positive and negative charge in a species leads to a dipolar situation. Although the electrical field may be symmetrical, transient dipoles induce dipoles in neighbouring ions, leading to induced dipole-dipole attractions, with a consequent enhancement of the total lattice energy. The presence of such enhancements may be revealed by a comparison of interionic distances with the sums of the corresponding ionic radii.

We return to the MX compounds because the MX_2 structures are well behaved with respect to the radius ratio criteria. Table 2.10 list the radii sums; $\Sigma_i r_i$ for AgI has been corrected to 4-coordination. Among ions of similar size, polarization effects are generally

Table 2.10 Interionic distances and radii sums

Compound	Structure type	r_e/nm	$\Sigma_i r_i$	Δ/nm
NaF	NaCl	0.231	0.231	0.00
NaCl	NaCl	0.282	0.282	0.00
NaBr	NaCl	0.298	0.299	0.01
NaI	NaCl	0.323	0.324	0.01
AgF	NaCl	0.246	0.246	0.00
AgCl	NaCl	0.277	0.297	0.20
AgBr	NaCl	0.288	0.314	0.26
AgI	β-ZnS	0.281	0.322	0.41

Table 2.11 Effective atomic numbers for Na⁺, K⁺ and Ag⁺ ions

	Na^+	K^+	Ag^+
Z_{eff}	7	10	8
Z_{eff}/Z	0.64	0.53	0.17

more marked when an outer shell of d electrons is present, probably because of the screening effect of the d electrons on the nuclear charge. The radius of the Ag^+ ion is similar to that of Na^+ or K^+. We compare the screening in these three ions in Table 2.11, using Slater's rules[16] to calculate the screening constants. The screening of the nuclear charge in the silver ion results in its electron distribution being more easily distorted (polarized) by its neighbours than that of ions of similar size which have the inert-gas electron configuration. A physical manifestation of this effect can be traced in the solubility relationships among these halides[7] (see also Section 2.3.6).

Inorganic layer and molecular MX₂ structures

As the radius ratio in the MX_2 structures decreases, or/and polarization increases, the species in the crystal tend first to segregate into composite layers. A common layer-structure type is that shown by cadmium iodide, which is found for many metallic halides, hydroxides and chalcides. Figure 2.10 shows the cadmium iodide structure type. It belongs to the hexagonal space group $P\bar{3}m1$ with one formula-entity per unit cell: Cd^{2+} is in the Wyckoff a position 0, 0, 0, and I^- in d positions 1/3, 2/3, z and 2/3, 1/3, $\bar{z}$.

The composite layer consists of cadmium ions coordinated octahedrally by six iodide ions, the iodide ions having three cadmium ions as nearest neighbours, such that I^- forms the apex of a pyramid with three Cd^{2+} ions at its basal corners. The axial ratio c/a is 1.613: the anions are very nearly close-packed hexagonal; if it were exactly close-packed, the ratio would have been 1.633.

The Cd—I bond distance is 0.299 nm, and the sum of the ionic radii is 3.13 nm. The significant decrease is an indication of the covalent character in cadmium iodide. The radius ration criterion is upheld, however, because of the dominance of the electrostatic component of the lattice energy, through the doubly-charged cation. Cadmium chloride

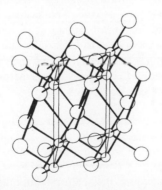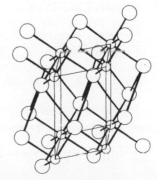

Fig. 2.10. Stereoview of the unit cell and environs of the cadmium iodide structure; the circles in order of decreasing size represent I^- and Cd^{2+}. The space group is $P\bar{3}m1$, with Cd^{2+} in Wyckoff a and I^- in d positions.

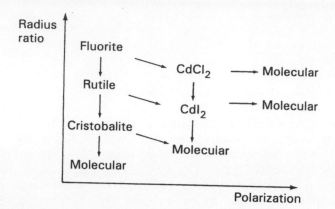

Fig. 2.11. Schematic transformations among MX_2 structure types according to radius ratio and polarization: $CdCl_2$ and CdI_2 are typical layer structures, intermediate in type between ionic and molecular compounds.

is also a layer-type structure, but the anions form a close-packed cubic array. The departure from additivity is only 0.05 nm; the polarizability of a chloride anion is less than one-half that of an iodide ion.

As the polarization and size factors lead further away from ionicity, molecular structures develop. A typical MX_2 molecular structure type is exhibited by mercury(II) chloride. The crystals of this compound are orthorhombic, space group *Pbnm*, with the atoms occupying three sets of Wyckoff c positions in the space group. The linear species ClHgCl are packed into the unit cell in a manner very similar to that of I_2 molecules in iodine (q.v.). The six nearest chlorine neighbours of mercury form a slightly distorted octahedral arrangement, and three mercury atoms and one chlorine atom coordinate to form a triangular pyramid. There are, thus, two Hg—Cl nearest neighbour distances, 0.223 nm and 0.227 nm, both of which are significantly less than the sum of the radii (0.291 nm).

Figure 2.11 shows a scheme for the transformations of structure types according to the radius ratio and polarization parameters. It is, of course, only approximate, and the actual structure obtained will always be determined by a minimum energy conformation in which the different types of interatomic forces all play a part. The situation at this stage highlights our previous statement that no classification of structures is completely satisfactory. For example, on the one hand, some authors tend to regard the zinc sulfide structures as covalent on account of the structural similarity between β-ZnS and diamond. On the other hand, the electrostatic model, expressed by (2.8), gives a satisfactory value for the lattice energy for this and similar compounds (see also Section 3.3).

2.3.5 Complex ion structures

Many structures exist in which the components, or structural units, are held together by forces that are essentially ionic, but in which some of the structural units themselves are polyatomic and involve different forces of attraction. Compounds such as potassium sulfate, sodium nitrate, calcium carbonate and potassium hexachloroplatinum(IV) fall into this class. The complex ion comprises a fragment in which its components are linked by mainly covalent bonds, but carries a residual charge that is balanced by the cations present in the structure. Sulfates contain the tetrahedral group $[SO_4]^{2-}$ of point-

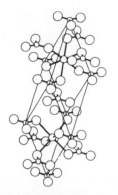 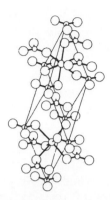

Fig. 2.12. Stereoview of the unit cell and environs of the calcite structure; the circles in order of decreasing size represent O, Ca^{2+} and C species. The space group is $R\bar{3}c$, with Ca^{2+} in Wyckoff b, C in a and O in e positions.

group symmetry $\bar{4}3m$, nitrates and carbonates contain the $[NO_3]^-$ and $[CO_3]^{2-}$ groups, respectively, of point-group symmetry $\bar{6}m2$, and the hexachloroplatinum(IV) ion $[PtCl_6]^{2-}$ is octahedral with symmetry $m3m$. Figure 2.13 shows the structure of the calcite form of calcium carbonate, which is isomorphous with sodium nitrate. These compounds crystallize with space group $R\bar{3}c$, which is one of the space groups that can be referred to hexagonal axes. The Ca^{2+} (Na^+) ions occupy Wyckoff positions b of symmetry $\bar{3}$, the C (N) atoms positions a of symmetry 32, and the oxygen atoms positions e with point-group symmetry 2. In Figure 2.12, the origin has been shifted from that given elsewhere[4] to the point ¼, ¼, ¼; we shall consider the effect of transformation to hexagonal axes by means of Example 2.1.

In the calcite structure, the planes of the carbonate (nitrate) ions are all normal to the direction of the (vertical) three-fold axis, the calcium (sodium) ions are coordinated octahedrally by six oxygen atoms from different carbonate (nitrate) ions, and each oxygen atom is linked to two calcium (sodium) species electrostatically and to one carbon (nitrogen) species covalently.

EXAMPLE 2.1. Calcite has been referred to a rhombohedral unit cell, space group $R\bar{3}c$, with $a = 0.6361$ nm and $\alpha = 46.12°$. The two formula-entities in the unit cell have the coordinates

2 C 0, 0, 0 ; ½, ½, ½.

2 Ca ¼, ¼, ¼ ; ¾, ¾, ¾.

6 O $x, \bar{x}, 0$; $0, x, \bar{x}$; $\bar{x}, 0, x$; ½ x, ½ + x, ½ ; ½, ½ - x, ½ + x , ½ + x, ½, ½ - x.

Determine the parameters of the triply-primitive hexagonal unit cell, and the atomic coordinates referred to that cell. The answers are at the end of the chapter.

Electrostatic valence rule

One of the general principles for ionic structures put forward empirically by Pauling stated that *in a stable structure the total electrostatic bond strength reaching an anion from all neighbouring cations is equal to the charge on the cation.* Put another way, *the charge on an ion tends to be neutralized by its nearest neighbours.* This rule is implicit in Evjen's method for the approximate calculation of a Madelung constant (see Section 2.2.1), which treats a block of structure as neutral, and we can see its application in the structures that we have considered as follows.

In applying the electrostatic valence rule, species are given their formal charge according to the particular structure, such as Na(+1) in $NaNO_3$, Cu(+2) in $CuSO_4$, Cl(-1) in NaCl, and so on. Then, the numerical value of that charge is divided by the number of electrostatic links to its *nearest* neighbours to give an **electrostatic bond strength**.

Thus, in cadmium iodide (Figure 2.10) the electrostatic bond strength ε of the Cd—I link is 1/3; hence, there will be three of these links to iodine to build up the numerical charge on that species to unity, as Figure 2.10 shows.

EXAMPLE 2.2. Is the electrostatic valence rule satisfied for the structure of sodium nitrate, Figure 2.12? The answer is at the end of the chapter.

2.3.6 Perovskites and spinels

Two structures of common occurrence are those of the perovskites, of general composition MNX_3, and spinels, of formula MN_2X_4. Normally, in perovskites, X is oxygen or fluorine, M is a large ion, such as K^+, Ca^{2+}, Ba^{2+} or Pb^{2+}, of size comparable to that of the X species and forming an approximately close-packed cubic array with them. The species N, such as Ti(IV), Ce(IV), Zr(IV) or Nb(V), are smaller in size and occupy the octahedral interstices (see Section 4.5.2 and Problem 4.13) in the close-packed $(M+X)$ array. Thus, each M species is coordinated by twelve X species, each N by six X, and each X by four M and two N; Figure 2.13 is a stereoview of the perovskite structure. In this ideal structure type, there is a relationship between the radii of the species. Thus,

$$r_M + r_X = \sqrt{2}(r_N + r_X).$$

In practice, however, there is a tolerance factor (Goldschmidt factor) g for the perovskite structures, such that

$$r_M + r_X = g\sqrt{2}(r_N + r_X),$$

where g ranges from approximately 0.7 to 1 (ideal). The size ranges found for M and N are 0.1–0.15 nm and 0.05–0.8 nm, respectively.

The electrostatic valence rule is satisfied by the perovskite structure type: the electrostatic bond strength M(II)–X is 1/6, and that of N(IV)–X is 2/3. Thus, the bond strength reaching the anion is numerically $(4 \times 1/6) + (2 \times 2/3)$, or 2, which is correct for X = oxygen.

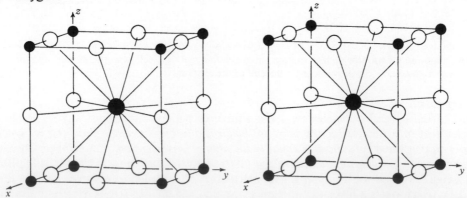

Fig. 2.13. Stereoview of the unit cell and environs of the ideal cubic perovskite structure type. The space group is $Pm3m$, with M at 0, 0, 0; N at ½, ½, ½; and O at ½, 0, 0; 0, ½, 0; and 0, 0, ½.

Often the perovskite structure is distorted in such a manner that the crystal structure ceases to be centrosymmetric and attains a permanent polarized state. Some polarized perovskites are **ferroelectric**: they resemble ferromagnets, but instead of the spins being aligned in *domains*, they are aligned over many unit cells which results in relative permittivities in the range 10^3 to 10^4. Thus, these materials have considerable applications in electronics. The absence of a centre of symmetry in perovskites leads to piezoelectricity in these materials, and $BaTiO_3$ and $NaNbO_3$ are important examples.

A typical ideal perovskite is strontium titanate $SrTiO_3$. It crystallizes with the cubic space group $Pm3m$, with one formula-entity per unit cell. The Sr atoms are in Wyckoff a positions, Ti in b, and O in c. The regular coordination polyhedron for titanium in this structure is a rhombic dodecahedron. It could be derived by drawing planes, normal to the Sr—O linkages, through the oxygen atoms, and allowing the planes to intersect. Another possible polyhedron is the combination of a cube and an octahedron; this solid is easily envisaged from Figure 2.13.

Spinels

A large number of oxides of the type MN_2O_4, such as $MgAl_2O_4$, forms the spinel structure type; other chalcogens or fluorine can replace oxygen, but they are less common. The structure is cubic, with 32 atoms per unit cell. Each M species is coordinated tetrahedrally by 4 oxygen atoms, and each N species octahedrally by 6 oxygen atoms; each oxygen atom is linked to one M and three N species. Thus, the structure shows a close-packed cubic arrangement of oxygen atoms, with M in the tetrahedral interstices, and N in the octahedral interstices of the close-packed array.

Generally, we find $M(II)$ and $N(III)$ species in spinels but, as with perovskites, overall electrical neutrality is the overriding factor, and other combinations of M and N species are possible, as with $MoNa_2O_4$, and $HgIn_2O_4$, for example.

Certain spinel structures, such as $MgFe_2O_4$ and $TiZn_2O_4$, retain the close-packed array of oxygen atoms, but the tetrahedral interstices are occupied by one-half of the N species, and the octahedral interstices are occupied at random by the other half of the N species together with the M species. These structures are known as **inverse spinels**, and may be formulated as $N(MN)O_4$.

The spinel structure is somewhat complex: Figure 2.14 is a plan of the unit cell with the halves along c shown separately; the numbers indicate the heights of the species in units of $c/8$. Calculations of lattice energies with the point-charge model indicate that the inverse spinel should be less stable than its normal counterpart, but this result may be modified by ligand-field stabilization[16] where N is a d-block metal.

One of the best known inverse spinels is the mineral *magnetite*, $Fe(III)[Fe(II)Fe(III)]O_4$. Inverse spinels often show electrical semi-conducting properties on application of an electric field, through an interchange of electrons among the species in the octahedral interstices.

We may define an **inversion factor** n for spinels such that $n = 0$ for a normal spinel MN_2O_4, and $n = 0.5$ for the an inverse spinel $N(MN)O_4$. Spinels exist with other values of n, particularly when the N species is d^0 or d^5 (zero ligand-field stabilization); examples are $Fe(III)[MnFe(III)]O_4$ with $n = 0.1$, and $Al[NiAl]O_4$ with $n = 0.4$.

The inverse spinels (*ferrites*) have interesting magnetic properties. Compounds containing atoms with unpaired electrons are **paramagnetic**; otherwise, they are **diamagnetic**. In an applied magnetic field, the electron spins align with the field, and the magnetic susceptibility κ follows the Curie law

$$\kappa = A + B/T,$$

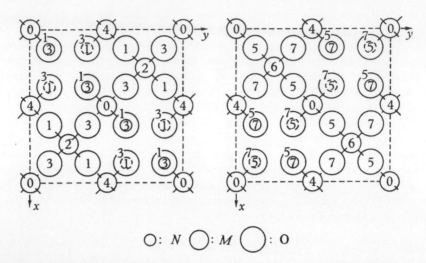

Fig. 2.14. Plan of the structure of a normal spinel MN_2O_4, as seen along z. The heights of the atoms along z are in units of $c/8$. The left-hand diagram is the half from $z = 0$ to $c/2$, and the right-hand diagram from $c/2$ to c.

where A and B are constants, and from this relationship the magnetic moment of the compound may be obtained.

In contrast, ferromagnetism and antiferromagnetism depend upon the cooperative action of many unit cells. In a **ferromagnetic**, the electron spins are aligned over 10^3–10^4 atoms in the presence of an applied magnetic field. If the aligned state is maintained below a cetrain transition temperature, the **Curie temperature**, the ferromagnetism is locked into the structure, and the magnetic moment is very high.

In an **antiferromagnetic**, the electron spins on adjacent atoms are antiparallel, so that the substance has a small magnetic moment; and example of antiferromagnetism is shown by the inverse spinel $Fe(III)[ZnFe(III)]O_4$.

Typically, normal spinels MN_2X_4 crystallize with the cubic space group $Fd3m$, with eight formula-entities per unit cell. The M species are in Wyckoff a positions, N in c and X in e.

2.3.7 Hydrates

Hydrates form a large class of mainly ionic compounds in which the Coulombic attraction between the structural units may be enhanced by hydrogen-bonding. The hydrogen bond is an electrostatic link to a second species, frequently, and most strongly, when that species is fluorine, oxygen, nitrogen, or chlorine and, to a lesser extent other species such as sulfur and phosphorus. In many hydrates, the water molecules surround the cation and serve to distribute its charge over a larger volume. This reduces its polarizing power and leads to ionic bonding between the coordinated cation and the anions in the structure. Thus, hydrates are common among compounds with small cations, such as Li^+, Na^+, Be^{2+}, Mg^{2+}, Al^{3+}, whereas corresponding compounds of the ions K^+, Rb^+, Ca^{2+} and Ba^{2+} are often anhydrous.

More specifically, in considering the structure of a hydrate it is useful to regard the molecule of water as a tetrahedron in which two of its apices, corresponding to the hydrogen atoms, are charged, say, $+1/6$, and two, corresponding to the lone pairs of electrons on oxygen, are charged $-1/6$. (Quantum mechanical calculations on the water molecule indicate that the charge on hydrogen is approximately $+0.16e$.) Thus, hydrates

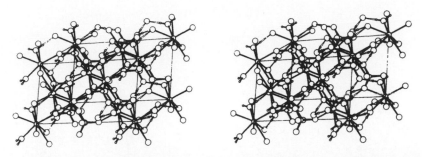

Fig. 2.15. Stereoview of the unit cell and environs of the calcium sulfate dihydrate (gypsum) structure, $CaSO_4 \cdot 2H_2O$. The circles in order of decreasing size represent O, Ca^{2+}, C and H species. The hydrogen bonds, shown by double lines (a hand lens may help) are responsible for cohesion of the structure normal to the y axis, the left-right direction in the illustration.

may be divided into classes.

There are those in which the water coordinates the cations, wholly or partially, those in which cations are not directly coordinated by water, the water molecules occupying interstitial positions in the structure, and there are hydrates in which water molecules have both of these functions; in some hydrate structures, chains of hydrogen-bonded water molecules exist.

In the first class, there are hydrates like $AlCl_3 \cdot 6H_2O$, or $[Al(H_2O)_6]^{3+} \cdot 3Cl^-$, and $CaSO_4 \cdot 2H_2O$. The existence of entities such as $[Al(H_2O)_6]^{3+}$ have been confirmed by the radial distribution curves[16] obtained by X-ray and neutron diffraction of aqueous solutions.

Interstitial hydrates are represented by zeolite silicate structures and clathrate compounds such as chloral hydrate $Cl_2 \cdot 8H_2O$. Copper sulfate pentahydrate shows two kinds of water in its structure. Four molecules are coordinated to the Cu^{2+} cation and one molecule of water is linked only to other water molecules and to oxygen atoms; hydrogen-bonding exists in this structure. In calcium sulfate dihydrate, gypsum, it is responsible for the cohesion of the structure in one direction.

Figure 2.15 illustrates the structure of gypsum; it is a layered structure. The sulfate ions are covalently-bonded entities of tetrahedral symmetry ($\bar{4}3m$). In the layer, each calcium ion is linked to six oxygen atoms of sulfate ions and to two oxygen atoms of water molecules. Each water molecule is linked to one calcium ion and to two oxygen atoms of different sulfate ions, one in its own layer and one in the adjacent layer. In these links, hydrogen-bonding has an important role: in fact, it is responsible for cohesion of the crystal between the layers. The softness and ease of **cleavage** of gypsum is related to these hydrogen-bonded layers; cleavage takes place across these layers, normal to the y axis, thus breaking hydrogen bonds, and the expansivity of the solid is significantly greater along this direction.

Ice

Although water is exceptional in many of its properties, it is not unreasonable to discuss ice at this stage. The chemically analogous compound H_2S adopts a close-packed cubic structure in the solid state, but ice has several, more complex structures. At 90 K ice adopts a hexagonal structure, similar to that of the tridymite form of silicon dioxide, with space group $P6_3/m\ mc$ ($P6_3mc$ with a centre of symmetry added). The oxygen atoms occupy the Wyckoff f positions, the sites of silicon in the tridymite form of silica. The intermolecular distance is 0.276 nm, so that the water molecule has an effective radius of

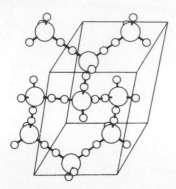 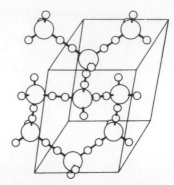

Fig. 2.16. Stereoview of the unit cell and environs of the hydrogen-bonded structure of ice at 90 K; the circles in order of decreasing size represent oxygen and half-hydrogen atoms. A tetrahedral disposition of bonds exists around any one oxygen atom, averaged over a finite crystal; only two of the four directions are actually occupied, and these in a random manner throughout the structure.

0.138 nm. Hydrogen atoms lie between two oxygen atoms, at two distances 0.101 nm and 0.137 nm. The larger of these two distances represents hydrogen-bonding between the molecules in ice.

At 90 K, the hydrogen-bond formation is nearly total. Figure 2.16 illustrates the structure of ice at 90 K. Each small circle represents a hydrogen atom site but only *two* out of every four such nearest-neighbour sites are occupied, and these in a statistical manner throughout the structure; each small circle may be thought of as a statistical half-hydrogen atom.

Spectroscopic and calorimetric measurements of the entropy of ice show that it possesses a *residual* entropy of 3.3 J K^{-1} mol^{-1}. There are six ways in which a tetrahedral structure may be orientated in space. Each adjacent H_2O molecule has two occupied and two unoccupied directions. The probability that one given direction is available for one hydrogen atom is 2/4, whereas for both atoms being located in the desired orientation the probability is $(2/4)^2$, or 1/4, since the two events are not correlated. Thus, the joint probability of a particular orientation of the given molecule in the ice structure $6 \times 1/4$. For a total of L molecules it is $(3/2)^L$. The entropy associated with this probability is, from the Boltzmann equation, $Lk_B \ln(3/2)$, or $R\ln(3/2)$, which is 3.37 J K^{-1} mol^{-1}, in very close agreement with the experimental result.

2.3.6 Aspects of solubility

Generally, an ionic solid MX is soluble in water when the free energy change on hydration of the ion-gas $M^+(g)$, $X^-(g)$ is more negative than the lattice free energy. This situation is illustrated by Figure 2.17, which has certain features in common with the right-hand side of Figure 2.2. The zero for free energies of both the crystal and the hydrated ions is the ideal ion-gas. The free energy of the crystal is the lattice enthalpy ΔH_1 (Figure 2.2) together with the entropic contribution $T\Delta S_1$, where ΔS_1 is given by

$$(2.20) \qquad\qquad \Delta S_1 = S(\text{cr}) - \Sigma S_{\text{ig}} ,$$

where $S(\text{cr})$ is the absolute entropy of the crystal relative to $S = 0$ for a perfect crystal at 0 K. The entropy of a gas can be calculated by statistical thermodynamics. For a monatomic gaseous species, the entropy is given by the Sackur-Tetrode equation[17]

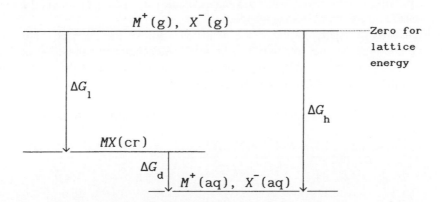

Fig. 2.17. Free-energy level diagram relating parameters of the thermodynamics of solubility.

$$(2.21) \qquad S = nR \ln[\exp(5/2) \, (2\pi m k_B T) h^2)^{3/2} \, (k_B T/p)] \,,$$

where n is the number of moles of the species of mass m and p is the pressure of the gas; the other symbols have their meanings as before.

The free energy of hydration of a gaseous ion is not easy to calculate, because of the complex interactions with the solvent water. However, the solubility in which we are interested is given through the free energy of dissolution ΔG_d. We write, for standard state conditions, that is, 1 atm and 25 °C,

$$(2.22) \qquad \Delta G_d^\circ = \Delta H_d^\circ - T\Delta S_d^\circ \,,$$

where ΔH_d° is the enthalpy change on dissolution to infinite dilution, obtained experimentally from calorimetric measurements. The entropy of dissolution is given by

$$(2.23) \qquad \Delta S_d^\circ = \Sigma S_{ig}^\circ - S^\circ(cr) \,,$$

where ΣS_{ig}° is the sum of the entropies of the hydrated ions relative to $S_{H+}^\circ(g) = $ zero. The entropy of the crystal is obtained from heat capacity measurements as

$$S^\circ(cr) = \int_0^{298.15} [C_p(cr)/T] dT \,.$$

Entropy data are readily available in the literature[16,26].

The standard states relevant to solubility are the ion-gas, which is ideal, the crystal, which is defined to have unit activity under standard state conditions, and the hydrated ion at infinite dilution. The standard state of infinite dilution corresponds with unit mean activity[17]. Thus, all the terms are on a common basis, and we can calculate ΔG_d° from the available experimental data.

We consider two series of compounds, the silver halides and the magnesium halides; their solubility data are listed in Table 2.12. Qualitatively, negative ΔG_d° values apply to those substance that we call 'soluble', whereas positive ΔG_d° values apply to 'insoluble' substances. We can see at once that arguments based on enthalpy, instead of free energy, can be misleading with small, highly-charged ions.

Ions are hydrated in solution, that is, each ion tends to surround itself with a number of water molecules in a **hydration sphere**, so that an entity $M(H_2O)_n$ exists ($n \approx 6$) in dynamic equilibrium with solvent water molecules. Thus, part of the structure in the solid hydrate tends to be preserved in aqueous solution.

A decrease (more negative) value of ΔG_h° tends to stabilize the hydrated state with respect to the ion-gas, so that solubility is promoted. A decrease in either ΔG_l° or ΔS_d° tends to decrease solubility, ΔG_l° by stabilizing the crystal with respect to the ion-gas, and ΔS_d° by making the hydrated state relatively less probable.

Table 2.12 Solubility parameters for the silver and magnesium halides

	ΔG_d°/kJ mol^{-1}	ΣS_i°/J mol^{-1}	S°(cr)/J mol^{-1}	$T\Delta S_d^\circ$/ kJ mol^{-1}	ΔG_d°/kJ mol^{-1}
AgF	−20.5	64.4	83.7	−5.8	−14.7
AgCl	66.5	129.3	96.2	9.9	56.6
AgBr	84.1	154.8	107.1	14.2	69.9
AgI	111.7	183.3	114.2	20.0	91.1
MgF$_2$	−18.4	−137.2	57.3	−58.0	39.6
MgCl$_2$	−155.2	−7.5	89.5	−28.9	−126.3
MgBr$_2$	−186.2	43.5	123.0	−23.7	−162.5
MgI$_2$	−214.2	100.4	145.6	−13.5	−200.7

The transfer of an ion from the gas phase to water involves two processes. A *structure-breaking* effect occurs because the interaction between the ion and water molecules causes a disruption of the hydrogen-bonded structure of the water. *Structure-making* arises because of the coordination of water molecules around the ion, forming the hydration sphere. Any given case involves the interplay of these factors, and a quantitative explanation in molecular terms is complex. However, the thermodynamic calculation is precise within the limits of error on the parameters.

Consider the equilibrium

$$MX(cr) + aq \Leftrightarrow M^+(aq) + X(aq) .$$

From standard thermodynamic arguments, the equilibrium constant K is given in terms of activities by

$$K = a_{M+(aq)} \, a_{X-(aq)}/a_{MX(cr)}$$
$$= a_{M+(aq)} \, a_{X-(aq)} ,$$

since the activity of the solid in the standard state is defined to be unity. This equation may be re-cast as

$$K = a_\pm^2 = (m/m^\circ)^2 \gamma_\pm^2 ,$$

where m is the molality at saturation (solubility), $m^\circ = 1$ mol kg^{-1}, and $\gamma_\pm$ is the mean activity coefficient. Since, generally,

$$\Delta G^\circ = -RT \ln(K) ,$$

it follows that

(2.24) $$\Delta G_d^\circ = -RT \ln[(m/m^\circ)^2 \gamma_\pm^2].$$

In the case of sodium chloride, which is one of the small number of substances for which activity data at saturation are known, following (2.22), with $\Delta H_d^\circ = 3.8$ kJ mol^{-1} and $\Delta S_d^\circ = -43.1$ J mol^{-1}, $\Delta G_d^\circ = -9.05$ kJ mol^{-1}. From (2.24), since $m = 6.1$ mol kg^{-1} and $\gamma_\pm = 1.02$ at that concentration, $\Delta G_d^\circ = -9.06$ kJ mol^{-1}, which is excellent agreement with the value from (2.22). Further discussions on solubility have been given in the literature[16].

2.5 PROPERTIES ASSOCIATED WITH IONIC SOLIDS

The ionic bond is non-directional in character, and ions are coordinated to varying numbers of neighbours among the known ionic structures. The actual structure of a given substance is governed to a first approximation by geometrical, packing considerations, subject to the requirement of electrical neutrality of the structure as a whole. Polarization and covalent character enhance the lattice energies of ionic compounds, and may modify geometrical predictions based on the radius ratio criterion. Molecules do not occur in ionic solids, because electrons are almost totally localized in the atomic orbitals of the component ions.

Hydrates contain molecules of water, but the main components of structure are ionic. In polyatomic species, such as nitrates and sulfates, the atoms in the polyatomic ions are covalently bonded, but they are held with the complementary ions by mainly Coulombic forces in forming the crystal.

Ionic solids form mechanically hard crystals of low compressibility and thermal expansivity, but with relatively high melting points. Normally, they are electrical insulators in the solid state, but in the melt or in solution conduct electricity by transport of ions under the influence of an applied potential difference. Certain solids in which disorder exists, such as $RbAg_4I_5$, exhibit high electrical conductivity, but this feature is related more to the presence of disorder than to the nature of the ionic components (see also Section 5.8).

The trends in physical properties can be related to lattice energies. For example, the variation in melting point T_m can be linked with that in the lattice energy for a series of compounds. In fact, for the sodium halides the relationship is very closely linear:

	NaF	NaCl	NaBr	NaI
T_m/K	1266	1074	1020	934
$U(r_e)/kJ\ mol^{-1}$	919	775	741	691

The suggestion that solubility is related to covalent character is not generally satisfactory, as we have shown by the analysis in Section 2.3.6. Furthermore, compounds such as magnesium and calcium oxides are nearly as ionic as sodium chloride, yet they do not dissolve in water. The reason for this fact is that their lattice energies lie in the region of 3500–4000 kJ mol^{-1}. The O^{2-} ion hydrates to a pair of OH$^-$ ions, and the free energy of hydration for this process is only about 1000 kJ mol^{-1}. Thus, the dissolution of these oxides in water is thermodynamically unfavourable, and they remain insoluble.

ANSWERS TO EXAMPLES 2

2.1. From (1.12) and following (1.8), we have
$$a_H^2 = a_H \cdot a_H = (a_R - b_R) \cdot (a_R - b_R) = a_R^2 + b_R^2 - 2a_R b_R \cos(\alpha) = 2a_R^2[1 - \cos(\alpha)],$$
so that $a_H (= b_H) = 0.4983$ nm. Similarly, remembering that $a_R = b_R = c_R$ and $\alpha = \beta = \gamma$,

$$c_H{}^2 = c_H \cdot c_H = (a_R + b_R + c_R) \cdot (a_R + b_R + c_R) = 3a_R{}^2[1 + 2\cos(\alpha)]^{\frac{1}{2}},$$

so that $c_H = 1.7020$ nm. Following (1.8) and (1.9), we confirm readily that $\alpha = \beta = 90°$ and $\gamma = 120°$. Fractional coordinates are inversely proportional to the unit cell lengths, so that we use (1.11) transposed[2] to transform x, y and z. Thus, $x_H = (2x_R - y_R - z_R)/3$, $y_H = (x_R + y_R - 2z_R)/3$ and $z_H = (x_R + y_R + z_R)/3$. The eight corners of the rhombohedral unit cell transform to give points such as 0, 0, 0; 2/3, 1/3, 1/3, and so on. Three of these, 0, 0, 0; 2/3, 1/3, 1/3; 1/3, 2/3, 2/3; lie within the triply-primitive hexagonal unit cell and are Bravais-lattice points.

For C: $x_C = [(2 \times 0) - 0 - 0]/3 = 0$, $y_C = 0 + 0 - (2 \times 0)]/3 = 0$, $z_C = (0 + 0 + 0)/3 = 0$;
$x_C = [(2 \times \frac{1}{2}) - \frac{1}{2} - \frac{1}{2}]/3 = 0$, $y_C = \frac{1}{2} + \frac{1}{2} - (2 \times \frac{1}{2})]/3 = 0$, $z_C = (\frac{1}{2} + \frac{1}{2} + \frac{1}{2})/3 = \frac{1}{2}$.

For Ca: $x_{Ca} = [(2 \times \frac{1}{4}) - \frac{1}{4} - \frac{1}{4}]/3 = 0$, $y_{Ca} = [\frac{1}{4} + \frac{1}{4} - (2 \times \frac{1}{4})]/3 = 0$, $z_{Ca} = (\frac{1}{4} + \frac{1}{4} + \frac{1}{4})/3 = \frac{1}{4}$;
$x_{Ca} = [(2 \times \frac{3}{4}) - \frac{3}{4} - \frac{3}{4}]/3 = 0$, $y_{Ca} = [\frac{3}{4} + \frac{3}{4} - (2 \times \frac{3}{4})]/3 = 0$, $z_{Ca} = (\frac{3}{4} + \frac{3}{4} + \frac{3}{4})/3 = \frac{3}{4}$.

For O: $x_O = (2x + x)/3 = x$, $y_O = (x - x)/3 = 0$, $z_O = (x - x)/3 = 0$, and so on. Finally, we write

(0, 0, 0; 2/3, 1/3, 1/3; 1/3, 2/3, 2/3)+

C   0, 0, 0; 0, 0, ½.

Ca 0, 0, ¼; 0, 0, ¾.

O x, 0, 0; 0, x, 0; $\bar{x}$, $\bar{x}$, 0; $\bar{x}$, 0, ½; 0, $\bar{x}$, ½; x, x, ½.

2.2. Each sodium ion is coordinated by six oxygen atoms of different sulfate groups, so that ε(Na–O) = 1/6. Each oxygen atom is bound to two sodium ions, so that the electrostatic bond strength reaching the nitrate ion from these linkages is 1/3. Since there are three of them, the totals is 1, which is numerically equal to the anionic charge. Thus, the rule is upheld in the sodium nitrate structure.

PROBLEMS 2

2.1. Consider a 3 × 3 array of equidistant, alternating positive and negative unit charges. Using Evjen's method, calculate the Madelung constant for the array. Repeat the calculation for squares twice (5 × 5) and three times (7 × 7) the initial size. Estimate the likely true value for this parameter.

2.2. The sulfate ion may be considered as a regular tetrahedral arrangement of oxygen atoms around sulfur. If the charge on each oxygen atom is +1e and the S—O distance is 0.130 nm, calculate the electrostatic self-energy of the sulfate ion.

2.3. If $r(Cs^+) = 0.184$ nm and $r(I^-) = 0.212$ nm, and the departure of r_e from strict additivity is -0.001 nm, calculate the density of cesium iodide (Cs = 132.9; I = 126.9).

2.4. Rutile TiO_2 crystallizes in the tetragonal system with space group $P4_2/m\ nm$ and two formula-entities in a unit cell of dimensions $a = 0.4593$ nm and $c = 0.2959$ nm. The titanium atoms occupy Wyckoff a positions (0, 0, 0; ½, ½, ½) and the oxygen atoms the f positions (x, x, 0; $\bar{x}$, $\bar{x}$, 0; ½ + x, ½ – x, ½; ½ – x, ½ + x, ½) with $x = 0.3056$. Calculate the two Ti—O bond lengths. (The Ti and O radii are ca 0.075 nm and 0.125 nm, respectively.)

2.5. Using the standard enthalpy data below, show how the formation of $MgCl_2(cr)$ is preferred to that of MgCl(cr) (ΔH_i and ΔH_e refer to 0 K):

Process	Enthalpy/kJ mol⁻¹
Sublimation of Mg(cr)	149.0
First ionization of Mg(g)	737.7
Second ionization of Mg^+(g)	1451
Dissociation of Cl_2(g)	243.0
Electron affination of Cl(g)	−348.6
Formation of MgCl(cr)	−221.8
Formation of $MgCl_2$(cr)	−641.8

2.6. Calcium oxide has the sodium chloride structure type, with $a = 0.4811$ nm; $\rho/r_e = 0.162$. From the data below determine the affinity of oxygen for two electrons.

Sublimation enthalpy of Ca(cr)	176.6 kJ mol^{-1}
First ionization energy of Ca(g)	589.5 kJ mol^{-1}
Second ionization energy of Ca$^+$(g)	1145.0 kJ mol^{-1}
Dissociation enthalpy of O$_2$(g)	489.9 kJ mol^{-1}
Enthalpy of formation of CaO(cr)	−635.5 kJ mol^{-1}

2.7. Show that the radius ratio for the wurtzite structure type with atoms in maximum contact is 0.225.

2.8. Two polymorphs of titanium dioxide have the following crystal parameters:

(a) Rutile   Tetragonal $a = 0.4593$ nm, $c = 0.2959$ nm

2 Ti 0, 0, 0; ½, ½, ½.

4 O $x, x, 0$; $\bar{x}, \bar{x}, 0$; ½ + x, ½ − x, ½; ½ − x, ½ + x, ½.

($x = 0.3056$)

(b) Anatase   Tetragonal $a = 0.3785$ nm $c = 0.9514$ nm

(½, ½, ½)+

4 Ti   0, 0, 0; 0, ½, ¼.

8 O   0, 0, x; 0, 0, $\bar{x}$; 0, ½. ¼ + x; 0, ½, ¼ − x;

($x = 0.2066$; $r_{e,min} = 0.1910$ nm)

Calculate the Madelung constants for these structures (see Appendix 2), and indicate the probable order for their stability

2.9. In beryllium sulfate tetrahydrate, BeSO$_4$.4H$_2$O, each beryllium atom is coordinated by four water molecules; each water molecule is linked to one beryllium atom and, by hydrogen bonds, to two oxygen atoms of two neighbouring sulfate groups; each oxygen atom of the sulfate group is linked to two water molecules. (a) Does this arrangement satisfy the electrostatic valence rule? (b) How might the structure be better formulated? (c) What is the function of the water molecules in this hydrate?

2.10. β-Ethan-1,2-diaminotetracarboxylic acid (β-*edta*) (CO$_2$H)$_2$NCH$_2$CH$_2$N(CO$_2$H)$_2$ is monoclinic, space group $C2/c$, with $a = 1.327$, $b = 0.5575$, $c = 1.611$ nm, $\beta = 96.26°$. Intermolecular hydrogen-bonding exists between O$_1$ of one molecule and O$_2$ of an adjacent molecule. The relevant atomic fractional coordinates are as follow:

	x	y	z
O$_1$	−0.0843	0.8263	−0.0489
O$_2$	−0.1677	1.2186	−0.0698
H	−0.130	1.006	−0.069

(a) Calculate the O$_1$...O$_2$ hydrogen-bonded distance. (b) Is the hydrogen bond symmetrical? If not, list the two O...H distances. (c) Is the hydrogen bond linear? If not, list the O$_1$...H...O$_2$ angle.

2.11. Is the electrostatic valence rule satisfied for the perovskites NaNbO$_3$ and KZnF$_3$, and for the inverse spinel, magnetite Fe$_3$O$_4$? What is a better formulation for magnetite?

2.12. The solubility of lithium fluoride in water at 25 °C is 0.09 mol kg^{-1}. Calculate ΔG_d°, the activity coefficient may be obtained from the Debye-Hückel limiting law in the form $\ln(\gamma_{\pm}) = -1.172\sqrt{m}$. Compare the result with that using $\Delta H_d^{\circ} = 4.6$ kJ mol^{-1}, $S^{\circ} = 36.1$ J mol^{-1}, $\Sigma S_{ig}^{\circ} = 31.5$ J mol^{-1}. If there is a discrepancy between the two results, comment on it.

2.13. The equation of state for a solid may be written as $(\partial U/\partial V)_T = -p + T\alpha/\kappa$, where α is the thermal expansivity and κ the isothermal compressibility. (a) Obtain a relationship for ρ/r_e ($\rho = -1/a$) of equation (2.5) and κ. (b) Neglecting the error in applying the result at 25 °C, verify equation (2.8) and calculate the lattice energy for sodium chloride, given a cell side of 0.564 nm and a compressibility of $4.1 \times 10^{-11} \, N^{-1} \, m^2$.

2.14. In connection with Figure 2.2, the ionization energy and electron affinity parameters are defined at 0 K. How is it that, at a finite temperature T, the terms $5RT/2$ per mole do not appear with ΔH_i° and ΔH_e°?

CHECKLIST 2

At the end of this chapter, you should be able to:

1. Calculate the number of unit cells in a crystal of given size;
2. Define electronegativity;
3. Define the electrostatic (Coulombic) energy for a pair of ions;
4. Define the Madelung constant and electrostatic energy for a crystal;
5. Calculate approximate Madelung constants for simple structures;
6. Define lattice energy (U) and recognize the point-charge model for U;
7. Calculate U with the point-charge model;
8. Calculate U from thermodynamic (experimental) parameters (Born-Haber cycle);
9. Explain the additivity of ionic radii and discuss the radii of individual ions;
10. Be familiar with simple structures of the type MX;
11. Define and calculate radius ratio for the MX structures considered;
12. Relate structure type to radius ratio and lattice energy;
13. Be familiar with simple structures of the type MX_2;
14. Describe polarization and its effect on ionic structures;
15. Appreciate the transition from ionic to layer-type and molecular structures in terms of polarization effects;
16. Give examples of complex-ion ionic structures, spinels and perovskites, and apply the electrostatic valence rule to them;
17. Define paramagnetism, diamagnetism, ferromagnetism and antiferromagnetism;
18. Discuss the role of water in hydrates and ice;
19. Discuss solubility in terms of thermodynamic parameters, and its relation to U;
20. Describe the structural and physical characteristics associated with the ionic bond.

3

Looking at covalent structures

3.1 INTRODUCTION

The nature of the classification of structures adopted in this book results in the covalent class being relatively small, although not unimportant. Solids in which the structural units are held together in the solid state by covalent forces need a three-dimensional disposition of covalent interatomic bonds, and we shall see how this can come about. The many compounds of organic chemistry are not included here: although covalent forces exist between the atoms in organic molecules, the molecules themselves are linked in the solid state by weaker intermolecular forces, so that organic solids will be considered under the heading of molecular solids.

It is not difficult to understand how charged species, such as Na^+ and Cl^- can attract one another, as we have discussed in the previous chapter. But how do two neutral hydrogen atoms unite to form a hydrogen molecule, and how do neutral carbon atoms bond to form the structure of diamond? We expect that the forces involved will be electrical, because such is the nature of matter. The qualitative concept of covalency is based on shared pairs of electrons but, further, we need to be able to explain features such as the near equivalence of bond lengths and angles involving given species across a range of compounds. We find the answer in the wave mechanics of the covalent bond, and we shall discuss this topic next.

3.2 WAVE MECHANICS OF THE COVALENT BOND

An electron that is free to move in a single dimension can be represented as a wave by the Schrödinger wave equation, which may be stated as

$$(3.1) \qquad [-(\hbar^2/2m_e)\,(d^2/dx^2) + V(x)]\,\psi(x) = E\psi(x) ,$$

where $\hbar$ ('cross-h') $= h/2\pi$, h being the Planck constant, m_e is the rest mass of an electron, $V(x)$ is the potential energy of the electron at a position x, $\psi(x)$ is the one-dimensional wavefunction, and E is the total energy, the sum of kinetic and potential energies. There is an infinite number of solutions to (3.1) for a free electron, $\psi(x) = \exp(ikx)$ and $\psi(x) = A \exp(ikx)$ being two of them, as can be shown by double differentiation. The wavefunction is interpreted as an amplitude, such that $|\psi(x)|^2 dx$ represents the probability of finding the electron in a region of space between x and $x + dx$.

The Heisenberg **uncertainty principle**, which derives from the Schrödinger equation, connects uncertainties in position δx and momentum δp_x by the expression

$$(3.2) \qquad \delta x \delta p_x \geq \hbar/2 .$$

Thus, the more precisely the position of an electron is defined, the less precisely known is its momentum, and *vice versa*; they cannot be specified simultaneously to a precision greater than $\hbar/2$.

If we generalize (3.1) to three dimensions and consider the electron to be in the hydrogen atom, we write

(3.3) $[-(\hbar^2/2\mu)\,\nabla^2 + V]\psi = E\psi$,

where μ is the **reduced mass**[16] of the electron–proton pair, given by $\mu = m_e m_p/(m_e + m_p)$, and ∇^2 is the Laplacian operator, given by $\nabla^2 = \partial^2/\partial x^2 + \partial^2/\partial y^2 + \partial^2/\partial z^2$; V, E and ψ have meanings as before. More concisely, we write

(3.4) $\mathbf{H}\psi = E\psi$,

where $\mathbf{H}$ is the Hamiltonian operator, $-(\hbar^2/2\mu)\nabla^2 + V$. The Schrödinger equation cannot be solved exactly for species with more than one electron, but approximate solutions enable useful results to be obtained, as we shall see.

The probability interpretation of the three-dimensional wavefunction given by Born is that the function $|\psi|^2\,d\tau$ represents the probability of finding an electron within the infinitesimal volume element $d\tau$, it implies that

(3.5) $N^2 \displaystyle\int_{-\infty}^{\infty} |\psi|^2\,d\tau = 1$,

where N is the **normalization constant** for the wavefunction ψ. A wavefunction must satisfy the conditions of being *finite*, *single-valued* and *continuous* in both itself and its derivative.

EXAMPLE 3.1. Calculate the reduced relative mass of the hydrogen bromide molecule, given the relative atomic masses H=1.0079 and Br = 79.904. The answer is at the end of the chapter.

3.2.1 Electron-in-a-box
Consider an electron confined to movement in one-dimensional box of length a under a potential V that is zero for $0 \le x \le a$, and infinite for $0 > x > a$. The wave equation is then (3.1) with $V = 0$, and the general solution[16] to this second-order differential equation is of the form

(3.6) $\psi = C\,\exp(ikx) + D\,\exp(-ikx)$,

where C and D here are constants, and $k = (2m_e E/\hbar^2)^{1/2}$. Using de Moivre's theorem, (3.6) can be expanded to give

(3.7) $\psi = B\,\cos(kx) + A\,\sin(kx)$,

where A and B here are constants. If the electron remains within the box, $\psi = 0$ at $x = 0$ and at $x = a$, so that $B = 0$ and $ka = n\pi$, where $n = 1, 2, 3 \ldots$. Using the expression for k, we obtain

(3.8) $\psi_n = A\,\sin(n\pi x/a)$,

where A, from (3.5) $= (2/a)^{1/2}$, so that

(3.9) $E_n = n^2\pi^2\hbar^2\,(2m_e a^2) = n^2 h^2/(8m_e a^2)$.

It is evident that the energy, translational energy in this case, is quantized in units determined by the quantum number n. Since $n \geq 1$, the minimum value of the energy, the **zero-point** energy, is $h^2/(8m_e a^2)$. A more detailed analysis[16-18] introduces quantization of vibrational, rotational and spin motion, so that an electron is specified fully by four quantum numbers: n (1, 2, 3, ...), the principal quantum number; l (0, 1, 2, ... $n-1$) and $m_l(l, l-1, l-2, ... -l)$, governing angular momentum; and s (½) for the electron spin.

EXAMLE 3.2. Evaluate the normalization constant A in equation (3.8). The answer is at the end of the chapter.

3.2.2 Atomic orbitals and the hydrogen atom m_l

The wavefunctions that are the solutions of (3.4) for the hydrogen atom provide the basis for a discussion of both atomic and molecular structure. Electronic wavefunctions in atoms and molecules are termed **orbitals**. The total wavefunction can be divided into two parts: thus,

$$(3.10) \qquad \psi_{n,l,ml} = R_{n,l}(r) Y_{l,m_l}(\theta,\phi)$$

where $R_{n,l}(r)$ represents the **radial** part, and $Y_{l,m_l}(\theta,\phi)$ the **angular** part, or **spherical harmonics,** of the total wavefunction. In this expression the Cartesian x, y and z coordinates have been replaced by the spherical r, θ and ϕ coordinates, whereupon the volume element $d\tau$ (= $dxdydz$) becomes[16] $r^2 \sin(\theta)\, dr\, d\theta\, d\phi$. The limits of the variables that cover all space are now $r = 1$ to ∞, $\theta = 0$ to π, $\phi = 0$ to 2π.

The radial part of (3.10) determines the *size* of the wavefunction, and the angular part determines its *shape*. The atomic orbitals $\psi_{n,0,0}$ are spherically symmetrical: they are termed s orbitals; they have no angular dependence, and decrease in amplitude exponentially with the distance r of the electron from the nucleus. Orbitals for which $l > 0$ (p, d, f, ...) have an important angular dependence, which has a bearing on the directional character of the covalent bond.

The probability of finding the electron of a hydrogen atom between distances r and $r + dr$ from the nucleus is given by $|\psi|^2 \times$ the volume enclosed by a spherical shell of radii r and $r + dr$, that is, $|\psi|^2 \times 4\pi r^2 dr$. The same volume element evolves from the integral

$$(3.11) \qquad r^2 dr \int_0^\pi \sin(\theta)\, d\theta \int_0^{2\pi} d\phi .$$

Figure 3.1 illustrates the functions R, R^2 and $4\pi r^2 R^2$ for $n = 1$ and $l = 0$ (1s atomic orbital), $n = 2$ and $l = 0$ (2s atomic orbital), and $n = 2$ and $l = 1$ (2p atomic orbital) plotted against r from 0 to 6 Å (0.6 nm). The maximum in $4\pi r^2 R^2$ for the 1s orbital occurs at 0.5292 Å, or at the Bohr radius a_0 for hydrogen:

$$(3.12) \qquad a_0 = 4\pi\varepsilon_0 h^2/(\mu e^2) .$$

The number of **nodes**, regions of zero amplitude and density, for a function based on $R_{n,l}$ is $n - 1$.

When the quantum number l is equal to or greater than unity, the density distribution is no longer spherically symmetrical. A useful simplification of the presentation of atomic orbitals may be achieved by defining three-dimensional surfaces that enclose a major fraction, say 0.95, of the total electron density. Thus, there is a 95% probability of

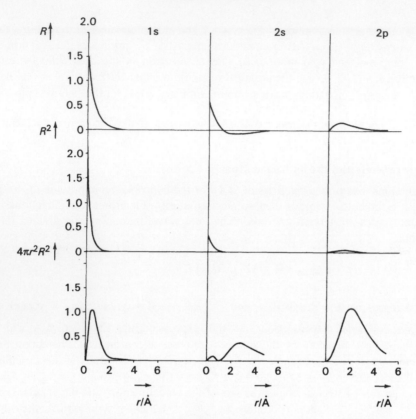

Fig. 3.1. The radial part $R_{n,l}$ of atomic orbitals as a function of distance $r/\text{Å}$ from the nucleus. For the 1s orbital a maximum occurs at the Bohr radius.

finding the electron in a one-electron species within the surface described.

The atomic orbitals for any given value of n comprise a **shell**, and the orbitals of varying l within a shell form a **sub-shell**. Thus, we obtain the common terminology for electrons in atoms:

n	1	2	3	4 ...
Shell	K	L	M	N ...
l	0	1	2	3 ...
Sub-shell	s	p	d	f ...

Figure 3.2 illustrates the shapes of the s, p and d atomic orbitals. The notation for p orbitals is straightforward: the p_x orbital has a nodal plane x,z. For the d orbitals, the lobes of d_{z^2} lie along the z axis, and those of $d_{x^2-y^2}$ lie along the x and y axes. The lobes of d_{xz}, d_{yz} and d_{xy} lie in the corresponding planes.

EXAMPLE 3.2. What is the probability that the 1s electron of hydrogen lies between $1.10a_0$ and $1.11a_0$ from the nucleus? The answer is at the end of the chapter.

3.2.3 Atoms with more than one electron
With multi-electron atoms, approximations must be made in obtaining a solution to the wave equation. Thus, if $\psi(1,2)$ is a wavefunction for the helium atom, we write it as a product of two one-electron wavefunctions:

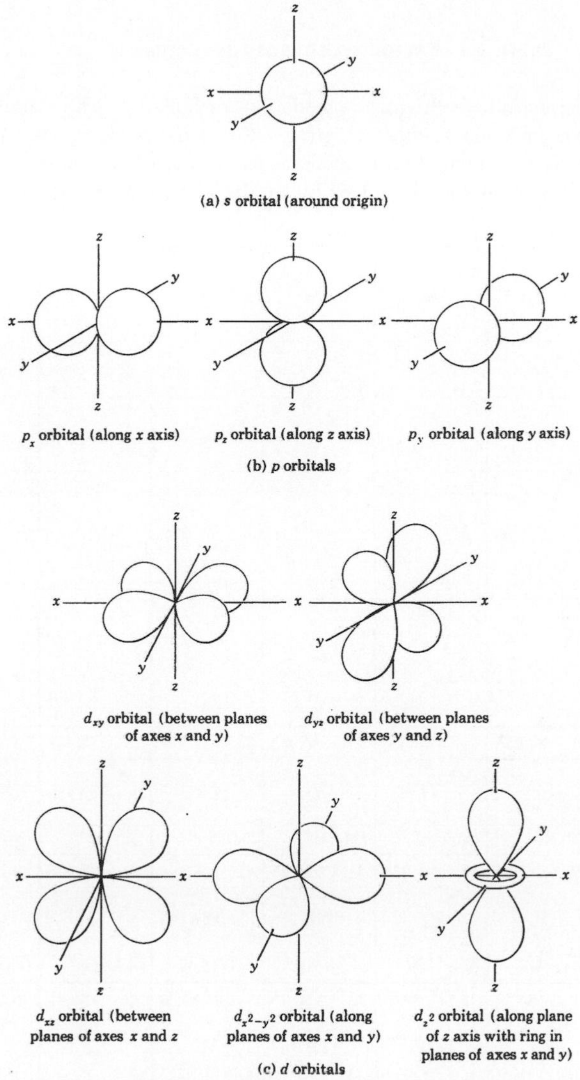

(a) *s* orbital (around origin)

p_x orbital (along *x* axis) p_z orbital (along *z* axis) p_y orbital (along *y* axis)

(b) *p* orbitals

d_{xy} orbital (between planes of axes *x* and *y*) d_{yz} orbital (between planes of axes *y* and *z*)

d_{xz} orbital (between planes of axes *x* and *z*) $d_{x^2-y^2}$ orbital (along planes of axes *x* and *y*) d_{z^2} orbital (along plane of *z* axis with ring in planes of axes *x* and *y*)

(c) *d* orbitals

Fig. 3.2. The angular parts of total (a) *s*, (b) *p*, and (c) *d* wavefunctions.

$$(3.13) \qquad \psi(1,2) = N\psi(1)\psi(2),$$

where N is the normalizing constant for the combined function. The nuclear charge in each one-electron function is modified in accordance with Slater's rules[16] to take account of both electrons, so that the atom number for the multi-electron atom is replaced by Z_{eff} (see also Section 2.3.1). Atoms with two or more electrons are subject to the **Pauli exclusion principle**, which allows an orbital to accommodate two electrons in the same quantum state (n, l, m_l) provided that their spins are paired, that is, when the spin angular momenta of the two electrons resolved along the *z* axis, often expressed by the quantum number m_s, are $\pm\frac{1}{2}$.

Aufbau principle
The atomic orbitals are intimately connected with the periodic table of the elements, Table 3.1. The sequence of energies of atomic orbitals is, normally,

Table 3.1 Periodic table of the elements

The periodic table numbering follows the recommendations of IUPAC[a]. The elements in groups 1 and 2, with helium, form the s block, those in groups 13–18 the p block, the lanthanides and actinides the f block, and the remaining, transition elements the d block. Groups 1–7 were formerly designated IA–VIIA, 8–10 were group VIII, 11—17 were IB–VIIB, and 18 was group 0. Each box

1	2	3	4	5	6	7	8	9
1 1.0079 **H** $(1s)^1$								
3 6.941(2) **Li** $(2s)^1$	4 9.0122 **Be** $(2s)^2$							
11 22.980 **Na** $(3s)^1$	12 24.305 **Mg** $(3s)^2$							
19 39.098 **K** $(4s)^1$	20 40.078(4) **Ca** $(4s)^2$	21 44.956 **Sc** $(3d)^1(4s)^2$	22 47.88(3) **Ti** $(3d)^2(4s)^2$	23 50.942 **V** $(3d)^3(4s)^2$	24 51.996 **Cr** $(3d)^5(4s)^1$	25 54.938 **Mn** $(3d)^5(4s)^2$	26 55.847(3) **Fe** $(3d)^6(4s)^2$	27 58.933 **Co** $(3d)^7(4s)^2$
37 85.468 **Rb** $(5s)^1$	38 87.62 **Sr** $(5s)^2$	39 88.906 **Y** $(4d)^1(5s)^2$	40 91.224 **Zr** $(4d)^2(5s)^2$	41 92.906 **Nb** $(4d)^4(5s)^1$	42 95.94 **Mo** $(4d)^5(5s)^1$	43 98.906 ^{99}**Tc** $(4d)^5(5s)^2$	44 101.07(2) **Ru** $(4d)^7(5s)^1$	45 102.91 **Rh** $(4d)^8(5s)^1$
55 132.91 **Cs** $(6s)^1$	56 137.33 **Ba** $(6s)^2$	71 174.97 **Lu** $(4f)^{14}(5d)^1(6s)^2$	72 178.49(2) **Hf** $(5d)^2(6s)^2$	73 180.95 **Ta** $(5d)^4(6s)^2$	74 183.85(3) **W** $(5d)^4(6s)^2$	75 186.21 **Re** $(5d)^4(6s)^2$	76 190.2 **Os** $(5d)^6(6s)^2$	77 192.22(3) **Ir** $(5d)^7(6s)^2$
87 223.02 ^{223}**Fr** $(7s)^1$	88 226.03 ^{226}**Ra** $(7s)^2$	103 262.11 ^{262}**Lr** $(5f)^{14}(6d)^1(7s)^2$	104 (260) **Ku** $(5f)^{14}(6d)^2(7s)^2$	105 (261) **Ha** $(5f)^{14}(6d)^3(7s)^2$				

57 138.91 **La** $(4f)^0(5d)^1(6s)^2$	58 140.12 **Ce** $(4f)^1(5d)^1(6s)^2$	59 140.91 **Pr** $(4f)^3(5d)^0(6s)^2$	60 144.24(3) **Nd** $(4f)^4(5d)^0(6s)^2$	61 146.92 ^{147}**Pm** $(4f)^5(5d)^0(6s)^2$	62 150.36(3) **Sm** $(4f)^6(5d)^0(6s)^2$	63 151.96 **Eu** $(4f)^7(5d)^0(6s)^2$
89 227.03 ^{227}**Ac** $(5f)^0(6d)^1(7s)^2$	90 232.04 **Th** $(5f)^0(6d)^2(7s)^2$	91 231.04 **Pa** $(5f)^2(6d)^1(7s)^2$	92 238.029 **U** $(5f)^3(6d)^1(7s)^2$	93 237.05 ^{237}**Np** $(5f)^4(6d)^1(7s)^2$	94 239.05 ^{239}**Pu** $(5f)^6(6d)^0(7s)^2$	95 241.06 ^{241}**Am** $(5f)^7(6d)^0(7s)^2$

[a] *Nomenclature of Inorganic Chemistry* (1989), Butterworth.
[b] *Pure & Applied Chemistry* Vol 63, pp. 987–988 (1991).

contains the chemical symbol of the element, its atomic number, relative atomic mass and outermost electronic configuration. The elements are arranged by group number and period (principal quantum number of outermost electron/s). The atomic masses are those recommended by IUPAC[b]; they are relative values, being scaled to $^{12}C = 12$. The precision is ± 1 in the last digit quoted, unless indicated otherwise.

10	11	12	13	14	15	16	17	18
								2　4.0026 **He** $(1s)^2$
			5　10.811(5) **B** $(2s)^2(2p)^1$	6　12.011 **C** $(2s)^2(2p)^2$	7　14.007 **N** $(2s)^2(2p)^3$	8　15.999 **O** $(2s)^2(2p)^4$	9　18.998 **F** $(2s)^2(2p)^5$	10　20.180 **Ne** $(2s)^2(2p)^6$
			13　26.982 **Al** $(3s)^2(3p)^1$	14　28.086 **Si** $(3s)^2(3p)^2$	15　30.974 **P** $(3s)^2(3p)^3$	16　32.066(6) **S** $(3s)^2(3p)^4$	17　35.453 **Cl** $(3s)^2(3p)^5$	18　39.948 **Ar** $(3s)^2(3p)^6$
28　58.963 **Ni** $(3d)^8(4s)^2$	29　63.546(3) **Cu** $(3d)^{10}(4s)^1$	30　65.39(2) **Zn** $(3d)^{10}(4s)^2$	31　69.723 **Ga** $(4s)^2(4p)^1$	32　72.61(2) **Ge** $(4s)^2(4p)^2$	33　74.922 **As** $(4s)^2(4p)^3$	34　78.96(3) **Se** $(4s)^2(4p)^4$	35　79.904 **Br** $(4s)^2(4p)^5$	36　83.80 **Kr** $(4s)^2(4p)^6$
46　106.42 **Pd** $(4d)^{10}(5s)^0$	47　107.87 **Ag** $(4d)^{10}(5s)^1$	48　112.41 **Cd** $(4d)^{10}(5s)^2$	49　114.82 **In** $(5s)^2(5p)^1$	50　118.71 **Sn** $(5s)^2(5p)^2$	51　121.76 **Sb** $(5s)^2(5p)^3$	52　127.60(3) **Te** $(5s)^2(5p)^4$	53　126.90 **I** $(5s)^2(5p)^5$	54　131.29(2) **Xe** $(5s)^2(5p)^6$
78　195.08(3) **Pt** $(5d)^9(6s)^1$	79　196.97 **Au** $(5d)^{10}(6s)^1$	80　200.59(2) **Hg** $(5d)^{10}(6s)^2$	81　204.38 **Tl** $(6s)^2(6p)^1$	82　207.2 **Pb** $(6s)^2(6p)^2$	83　208.98 **Bi** $(6s)^2(6p)^3$	84　209.98 ^{210}**Po** $(6s)^2(6p)^4$	85　209.99 ^{210}**At** $(6s)^2(6p)^5$	86　222.02 ^{222}**Rn** $(6s)^2(6p)^6$

64　157.25(3) **Gd** $(4f)^7(5d)^1(6s)^2$	65　158.93 **Tb** $(4f)^9(5d)^0(6s)^2$	66　162.50(3) **Dy** $(4f)^{10}(5d)^0(6s)^2$	67　164.93 **Ho** $(4f)^{11}(5d)^0(6s)^2$	68　167.26(3) **Er** $(4f)^{12}(5d)^0(6s)^2$	69　168.93 **Tm** $(4f)^{13}(5d)^0(6s)^2$	70　173.04(3) **Yb** $(4f)^{14}(5d)^0(6s)^2$	**Lanthanides**
96　244.06 ^{244}**Cm** $(5f)^7(6d)^1(7s)^2$	97　249.08 ^{249}**Bk** $(5f)^9(6d)^0(7s)^2$	98　252.08 ^{252}**Cf** $(5f)^{10}(6d)^0(7s)^2$	99　252.08 ^{252}**Es** $(5f)^{11}(6d)^0(7s)^2$	100　257.10 ^{257}**Fm** $(5f)^{12}(6d)^0(7s)^2$	101　259.10 ^{258}**Md** $(5f)^{13}(6d)^0(7s)^2$	102　259.10 ^{259}**No** $(5f)^{14}(6d)^0(7s)^2$	**Actinides**

$$1s < 2s < 2p < 3s < 3d < 4s \ldots .$$

Electrons are fed into the orbitals in the *Aufbau* order $(1s)^1$, $(1s)^2$, $(2s)^1$, $(2s)^2$, $(2p)^1$ for hydrogen, helium, lithium, beryllium and boron. Then, with carbon, we have to consider whether the next electron enters the same or a different p orbital. The dilemma is resolved by **Hund's rule** which states that, in the lowest energy state, or **ground state**, orbitals are occupied singly as far as possible.

For atomic numbers greater than 20 ($n \geq 3$), the nd orbitals have energies lying between the $(n + 1)s$ and $(n + 1)p$, and the *Aufbau* principle is followed in terms of these energies. Thus, the first transition series of elements have the configurations [Ar] $(3d)^1$ $(4s)^2$ at scandium to [Ar] $(3d)^{10}$ $(4s)^1$ at copper, where[Ar] represents the electron configuration of argon.

3.2.4 Variation method

In a multi-electron species, each electron is influenced by the potential fields of the nucleus and all other electrons present. In the hydrogen molecule, for example, six such interactions occur, Figure 3.3. It is these interelectronic attractions that make the solution of the wave equation generally intractable.

In approaching approximate solutions for the wave equation, we first assume the **Born-Oppenheimer approximation**: since the mass of a nucleus is very much heavier (*ca* 1840 times) than that of an electron, the motion of a nucleus is sluggish. Hence, the nuclei may be regarded as stationary while the electrons are free to move throughout the species. The kinetic energies of the nucleus and of the electrons are thus separated, and the electronic energy calculated separately as a function of the internuclear distance r. The total energy is then the sum of the electronic energy E and the repulsion energy of the nuclei.

We begin our introduction to the variation method by applying it to the hydrogen atom. We assume that the curve of energy against distance has a minimum at the equilibrium internuclear distance r_e, Figure 3.4. In the case of the hydrogen atom, we can identify the energy minimum with the ionization energy, whereas for a diatomic molecule the minimum energy corresponds to the theoretical bond dissociation energy D_e, which is the experimental quantity D_0 plus the zero-point energy of vibration.

In applying the variation method to the hydrogen atom, we write (3.4) as

(3.14) $$H\psi = E\psi,$$

where H is the *electronic* Hamiltonian, multiply both side by ψ and integrate over the space of the variables. Thus. we obtain

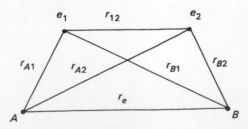

Fig. 3.3. Schematic diagram of the particle interactions in the hydrogen molecule: the nuclei are A and B, and the electrons e_1 and e_2; r_e is the equilibrium, or minimum energy, internuclear distance.

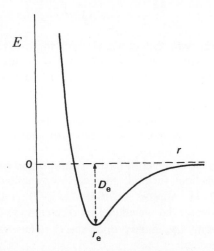

Fig. 3.4 Variation of electronic energy E with internuclear distance r in a diatomic molecule: the minimum theoretical dissociation energy D_e differs from the experimental quantity D_0 by the zero-point energy.

$$(3.15) \qquad E = \int \psi H \psi \, d\tau \, / \int \psi^2 d\tau \, .$$

We note that the Hamiltonian H is an operator acting on ψ, and the order of operations cannot be changed.

We adopt a trial function ψ_t, which we shall assume has the form $\exp(-ar)$, where a is a constant. Since this function is independent of θ and ϕ, we need only the radial part of the Hamiltonian which, in extenso in spherical coordinates, takes the form[16]

$$-[\hbar^2/(2\mu)][\partial^2 \psi_t/\partial r^2 + (2/r)\partial \psi_t/\partial r] - [e^2/(4\pi\varepsilon_0 r)]\psi_t =$$
$$\{[-\hbar^2/(2\mu)](a^2 - 2a/r) - [e^2/(4\pi\varepsilon_0 r)]\}\exp(-ar) \, ,$$

so that from (3.15)

$$(3.16) \qquad E = \frac{\int_0^\infty \{-[\hbar^2/(2\mu)](a^2 - 2a/r) - [e^2/(4\pi\varepsilon_0 r)]\} \exp(-ar) \, r^2 \, dr}{\int_0^\infty \exp(-2ar) \, r^2 \, dr} \, .$$

By use of the reduction formula (see Example 3.2) or the Gamma function[16], it is straightforward to show that (3.16) reduces to

$$(3.17) \qquad E = \hbar^2 a/(2\mu) - e^2 a/(4\pi\varepsilon_0) \, .$$

If we now differentiate E with respect to a and set the derivative to zero, we obtain

$$(3.18) \qquad a = \mu e^2/(4\pi\varepsilon_0 \hbar^2) \, ,$$

which, by comparison with (3.12) is the reciprocal of the Bohr radius. Hence, from (3.17) and (3.18),

(3.19) $E = -\mu e^4/(32\pi^2 \hbar^2 \varepsilon_0^2)$.

This result is identical with that obtained from the exact solution of the Schrödinger equation for the hydrogen atom[18], with $n = 1$. We made a very careful (or cunning) choice of the trial wave function, but the result of the analysis gives confidence in the variation method.

EXAMPLE 3.3. Use (3.19) to obtain the ground state energy in J and ionization energy in kJ.mol^{-1} for the hydrogen atom. The answers are at the end of the chapter.

3.2.5 Wavefunctions for molecules: linear combination of atomic orbitals
In setting up wavefunctions for molecules we may employ the technique known as the **linear combination of atomic orbitals** (LCAO), rather than guessing a trial wavefunction. Let a molecular wavefunction ϕ be composed as the sum of a **basis set** of normalized atomic orbitals ψ_n in the proportions c_n. Thus, we have

(3.20) $\phi = c_1 \psi_1 + c_2 \psi_2 + \ldots + c_n \psi_n = \sum_i c_i \psi_i$,

where the c_i are chosen so as to minimize the energy E, that is, $\partial E/\partial c_i = 0$ ($i = 1, 2, \ldots n$). For convenience, and without loss of generality, we let $n = 2$ for the substitution of (3.20) in (3.15). Because the Hamiltonian operator is **hermitian**[19], we know that $\int \psi_i^* H \psi_j \, d\tau = \int (\psi_j^* H \psi_i)^* \, d\tau$, and because we shall be considering only real functions, we write $\int \psi_i H \psi_j \, d\tau = \int \psi_j H \psi_i \, d\tau$. Then, we obtain

(3.21) $E = \dfrac{c_1^2 \int \psi_1 H \psi_1 d\tau + 2c_1 c_2 \int \psi_1 H \psi_2 d\tau + c_2^2 \int \psi_2 H \psi_2 d\tau}{c_1^2 \int \psi_1^2 d\tau + 2c_1 c_2 \int \psi_1 \psi_2 d\tau + c_2^2 \int \psi_2^2 d\tau}$.

If we write $\int \psi_i H \psi_j \, d\tau = H_{ij}$, and $\int \psi_i \psi_j \, d\tau = S_{ij}$, then

(3.22) $E = \dfrac{c_1^2 H_{11} + 2c_1 c_2 H_{12} + c_2^2 H_{22}}{c_1^2 S_{11} + 2c_1 c_2 S_{12} + c_2^2 S_{22}}$.

Since the atomic orbitals were chosen to be separately normalized, $\int \psi_i \psi_j \, d\tau \le 1$, the equality sign applying when $i = j$. It is a straightforward manipulation to form $\partial E/\partial c_1$ and $\partial E/\partial c_2$, and equate these derivatives to zero, whereupon we obtain the **secular equations**

(3.23) $c_1(H_{11} - ES_{11}) + c_2(H_{12} - ES_{12}) = 0$,

(3.24) $c_1(H_{12} - ES_{12}) + c_2(H_{22} - ES_{22}) = 0$,

and a solution for which not all of the c_n coefficients are zero requires that the **secular determinant** is equal to zero:

(3.25) $\begin{vmatrix} H_{11} - ES_{11} & H_{12} - ES_{12} \\ H_{12} - ES_{12} & H_{22} - ES_{22} \end{vmatrix} = 0$.

This result can be generalized to a system of n linearly combined atomic orbitals. The secular determinant is solved for E, which is then substituted back into the secular equations and solved for the c_n coefficients, incorporating the normalization condition

$$(3.26) \qquad\qquad \sum_i c_i^2 = 1 .$$

Orthogonality
Orthogonality in two functions ψ_i and ψ_j is defined by

$$(3.27) \qquad\qquad \int \psi_i \psi_j \, d\tau = \delta_{ij} ,$$

where δ_{ij} is the Kronecker delta; $\delta_{ij} = 1$ for $i = j$ but zero otherwise. The **overlap integral** S_{ij} measures the lack of orthogonality of the i,j pair of orbitals. In bonding situations, S_{ij} ranges between 0.2 and 0.3; qualitatively, it may be considered as a measure of the overlap of orbitals. Figure 3.5 illustrates several different degrees of overlap between atomic orbitals; the greater the overlap, the stronger is the bonding.

The **Coulomb integral** H_{ii}, or α, measures the energy of an electron when it occupies its own orbital. The **resonance integral**, H_{ij} ($i \neq j$), or β, is negative in a bonding situation, but vanishes to zero in the absence of bonding overlap.

EXAMPLE 3.4. If $f_1 = \cos(\theta)$ and $f_2 - \sin(\theta)$, is the function $\int f_1 f_2 \, d\theta$ orthogonal over the range $-\pi$ to $+\pi$. Find the normalization constants for the two functions. The answers are at the end of the chapter.

3.2.6 Molecular orbitals

We can examine a number of features of molecular orbitals by considering the simplest of species, the hydrogen-molecule ion H_2^+. We let ψ_1 and ψ_2 be normalized one-electron atomic orbitals for the electron in the neighbourhood of nuclei 1 and 2, respectively. We write a molecular wavefunction, molecular orbital, as

$$(3.28) \qquad\qquad \phi = c_1 \psi_1 + c_2 \psi_2 .$$

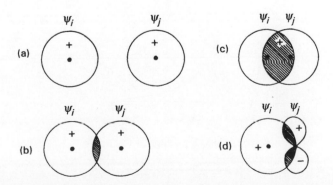

Fig. 3.5. The overlap of atomic orbitals ψ_i and ψ_j. (a) Two separated $1s$ atomic orbitals; $S = 0$. (b) Two $1s$ atomic orbitals with a small degree of overlap. (c) Two $1s$ atomic orbitals with strong overlap. (d) Overlap of $1s$ and $2p$; in this orientation $S = 0$ because of the cancelling property of the $\pm$ regions of the $2p$ atomic orbital in this orientation.

The two atomic orbitals ψ_1 and ψ_2 are equivalent, by symmetry, and probabilities are proportional to $|\phi|^2$, so that it follows that $c_1{}^2 = c_2{}^2$, or $c_2 = \pm c_1$. Following the discussion from (3.20), we obtain readily the secular equations

$$(3.29) \qquad\qquad c_1(\alpha - E) + c_2(\beta - ES) = 0 ,$$

$$(3.30) \qquad\qquad c_1(\beta - ES) + c_2(\alpha - E) = 0 ,$$

and the secular determinant

$$(3.31) \qquad\qquad \begin{vmatrix} \alpha - E & \beta - ES \\ \beta - ES & \alpha - E \end{vmatrix} = 0 .$$

Expanding the determinant and solving the quadratic in E gives two values,

$$(3.32) \qquad\qquad E_+ = (\alpha + \beta)/(1 + S) ,$$

$$(3.33) \qquad\qquad E_- = (\alpha - \beta)/(1 - S) ,$$

where α and β have been defined above, and S is implicitly S_{12}. Substituting the values of E in (3.29) and (3.30) shows that $c_1 = \pm c_2 = c_\pm$. From (3.28), we can write the two molecular orbitals as

$$(3.34) \qquad\qquad \phi_\pm = c_\pm(\psi_1 \pm \psi_2) .$$

Following (3.5),

$$(3.35) \qquad\qquad 1 = c_\pm{}^2 \int (\psi_1{}^2 + \psi_1{}^2 \pm 2\psi_1\psi_2)\, d\tau = c_\pm{}^2 (1 + 1 \pm 2S) ,$$

so that $c_\pm$ is $1/(2 \pm 2S)^{\frac{1}{2}}$. The value for S is[16] at the equilibrium internuclear distance of 0.16 nm is 0.56, so that we can identify two wavefunctions ϕ_+ and ϕ_-, corresponding to E_+

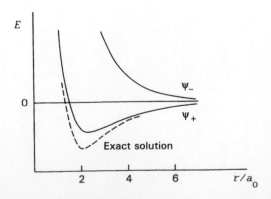

Fig. 3.6. Variation of energy E with internuclear distance r (in units of the Bohr radius a_0) for the bonding and antibonding molecular orbitals of the $H_2{}^+$ ion. The energy of the antibonding orbital approaches zero asymptotically.

and E. in (3.30) and (3.31). respectively:

(3.36) $\phi_+ = 0.566(\psi_1 + \psi_2)$,

(3.37) $\phi_- = 1.066(\psi_1 - \psi_2)$.

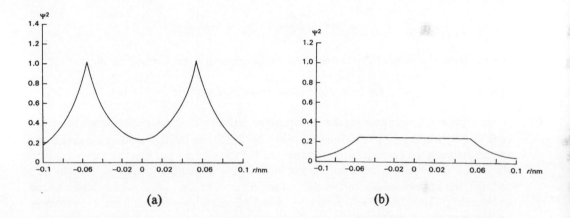

(a)   (b)

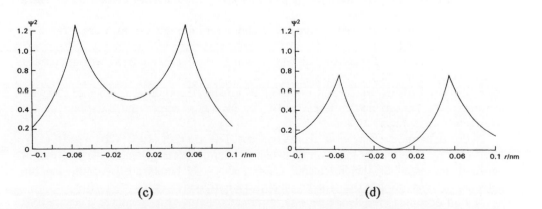

(c)   (d)

Fig. 3.7. Probability density functions for the H_2^+ species. (a) Superposition of ψ_1^2 and ψ_2^2. (b) The function $2\psi_1\psi_2$, which is constant for $|r_1 + r_2| \le r_e$. (c) The function $\psi_1^2 + \psi_2^2 + 2\psi_1\psi_2$, the superposition of (a) and (b), leading to a σ bonding molecular orbital. (d) The function $\psi_1^2 + \psi_2^2 - 2\psi_1\psi_2$ leads to a σ* antibonding orbital; the density in the internuclear region is depressed compared to that in (a).

Bonding and antibonding orbitals
The molecular orbital defined by (3.36) is a **bonding** molecular orbital, whereas (3.37) is an **antibonding** molecular orbital. Both of these molecular orbitals have cylindrical symmetry with respect to the internuclear axis, and are termed σ orbitals. Because both α and β are negative quantities. (3.36) represents the molecular (bonding) orbital of lower energy. Figure 3.6 is a plot of the energies of the bonding (σ) and antibonding (σ^*) orbitals as functions of internuclear distance r. The exact energy will always be lower than that calculated by an amount that will depend upon the level of approximation in the calculation.

Electron density
The probability density is determined by the square of the wavefunction. For the s bonding orbital we have

(3.38) $$\phi_+^2 = 0.566^2(\psi_1^2 + \psi_2^2 + 2\psi_1\psi_2).$$

If we assume the form of the $1s$ one-electron wavefunction, $\exp(-r/a_0)$, we obtain

(3.39) $$\phi_+^2 \propto \exp(-2r_1/a_0) + \exp(-2r_2/a_0) + 2\exp[-(r1 + r2)/a_0].$$

In Figure 3.7, we show (a) the superposition $\psi_1^2 + \psi_2^2$ at the equilibrium distance r_e of 0.11 nm, where $r_1 = |r + r_e/2|$ and $r_2 = |r - r_e/2|$; (b) the overlap density $2\psi_1\psi_2$ in the region between the two nuclei; (c) the sum of the three terms $\psi_1^2 + \psi_2^2 + 2\psi_1\psi_2$, which shows the enhancement of density in the internuclear region compared to that in (a); and (d) the superposition $\psi_1^2 + \psi_2^2 - (\psi_1^2 + \psi_2^2 + 2\psi_1\psi_2)$, which leads to the antibonding situation. The subtraction of the term $2\psi_1\psi_2$ constitutes a destructive interference which decreases the density compared with that of the sum of two separate atoms; in fact, the density falls to zero in the internuclear region

Orbital symmetry
The molecular orbitals of diatomic species exhibit symmetry, and this property forms a method by which they, and the bonding that they give rise to, may be classified. We have here three types of symmetry:
(a) Molecular orbitals that are symmetrical about the internuclear axis are termed σ-**orbitals**, Figure 3.8a–d;
(b) Molecular orbitals that have a nodal plane containing the internuclear axis are termed π-**orbitals**, Figure 3.8e,f;
(c) Molecular orbitals are termed **even** (*g*) or **odd** (*o*) according to whether they are symmetric or antisymmetric, respectively, for inversion across the mid-point of the internuclear axis, Figure 3.8a–f.

In this notation, we may write H_2 as $(1\sigma)^2$, and NO as $(1\sigma)^2 (1\sigma^*)^2 (2\sigma)^2 (2\sigma^*)^2 (1\pi)^4 (3\sigma)^2 (1\pi^*)^1$. The terminology can be extended usefully to larger molecules, so that in benzene, for example, the interplanar C—C and C—H bonds are σ-bonds, and the delocalized bonds formed from the $2p_z$ electrons are π-bonds.

Not all occupied orbitals are necessarily directly involved in bonding. Thus, in CO, which we may write as $(1\sigma)^2 (1\sigma^*)^2 (2\sigma)^2 (2\sigma^*)^2 (1\pi)^4 (3\sigma)^2$, the 1σ, 2σ and 1π molecular orbitals are bonding, $1\sigma^*$ and $2\sigma^*$ are antibonding; the electrons in the 3σ molecular orbital, which are directed away from each other, form a **non-bonding** molecular orbital.

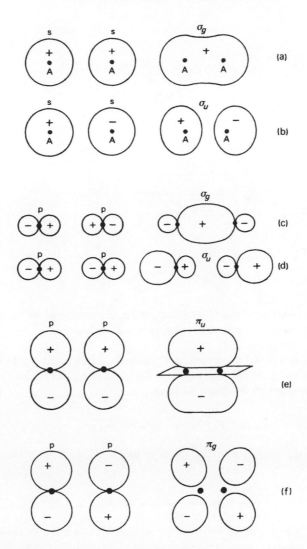

Fig. 3.8. Schematic diagrams of diatomic LCAO molecular orbitals. Bonding molecular orbitals are σ_g and π_u, whereas antibonding molecular orbitals are σ_u and π_g. For heteronuclear species σ_u is replaced by σ^*, and π_g by π^*.

3.2.7 Hybrid orbitals

It is evident from Figure 3.2 that, if bonds are to be formed by an overlapping of atomic orbitals, then, for significant overlap to occur, a strong directional character will arise in the bonding. However, we can see that this directional character will not immediately accord with experiment. Thus, if the water molecule is formed by an overlap of the $2p$ atomic orbitals of oxygen and the $1s$ atomic orbitals of hydrogen, we would expect an H–O–H bond angle of 90°, whereas it known from experiment to be 104.4°. Again, we need to obtain *four identical* C—H bonds in methane, with H–C–H angles of $\cos^{-1}(-1/3)$, or 109.47°.

In order to provide the necessary basis for bonding in these and similar compounds, we can introduce the model of the **hybrid** bond orbital. It is not necessarily a description of reality, but a device whereby we can explain the observed properties in terms of the theory developed so far.

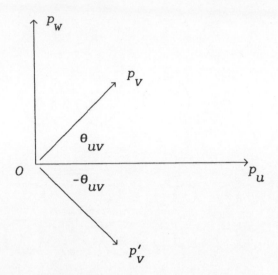

Fig. 3.9. Schematic formation of hybrid orbitals: p_u and p_w are orthogonal atomic orbitals, and p_v, a similar p orbital, lies at θ_{uv} to p_u, and in the same plane as p_u and p_w.

Consider two normalized p orbitals, say, p_u and p_v, at an angle θ_{uv} to each other, and let another normalized p orbital, say, p_w, lie in the plane of p_u and p_v, and be orthogonal to p_u, as illustrated by Figure 3.8. Then, we have

(3.40) $p_v = p_u \cos(\theta_{uv}) + p_w \sin(\theta_{uv})$.

Another normalized p orbital, similar to p_v, at an angle $-\theta_{uv}$ to p_u, may be termed $p_v{}'$ and be given by

(3.41) $p_v{}' = p_u \cos(\theta_{uv}) - p_w \sin(\theta_{uv})$.

We postulate two hybrid orbitals h and h', constructed from s and p orbitals according to

(3.42) $h = c_s s + c_p p_v$

and

(3.43) $h' = c_s s + c_p p_v{}'$.

In order to find expressions for c_s and c_p, we apply the normalization and orthogonality criteria to h and h'. From normalization, we have

(3.44) $1 = \int h^2 \, d\tau = c_s^2 \int s^2 \, d\tau + c_p^2 \int p_v^2 \, d\tau + 2c_s c_p \int sp_v \, d\tau = c_s^2 + c_p^2$;

a similar result arises for h'. By orthogonality:

(3.45) $\int hh' \, d\tau = \int \{c_s s^2 + c_p [p_u \cos(\theta_{uv}) + p_w \sin(\theta_{uv})][p_u \cos(\theta_{uv}) - p_w \sin(\theta_{uv})]\} \, d\tau$.

Since the sp and $p_u p_u$ integrals are zero, we find, from (3.44) and (3.45), that

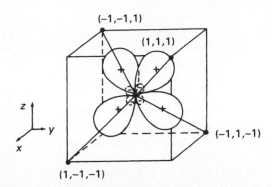

Fig. 3.10. Hybrid orbitals (sp^3) on carbon, showing their directional relationship to a cube. The axes of the orbitals lie along the three-fold symmetry axes of the cube. The relative coordinates at the marked corners refer to the carbon atom at the centre as origin.

$$(3.46) \qquad c_s^2 = \cos(2\theta_{uv})[\cos(\theta_{uv}) - 1], \quad c_p^2 = 1/[1 - \cos(2\theta_{uv})],$$

$$(3.47) \qquad c_s^2/c_p^2 = -\cos(2\theta_{uv}).$$

For some common values of $2\theta_{uv}$, we have the descriptions listed below. A diagram for the sp^3 hybrids of carbon, as in methane and diamond, is shown by Figure 3.10. The four labelled corners may be regarded as the sites of the bonded species, and the coordinates refer to the carbon atom at the centre as an origin.

$2\theta_{uv}$/deg	90	109.47	120	180
c_s^2	1	¼	1/3	½
c_p^2	0	¾	2/3	½
Type	p	sp^3	sp^2	sp

We have now reached a position where we can begin to discuss covalent solids, and to see how it is that so few solids are placed in this class under the chosen method of their classification.

3.3 COVALENT SOLIDS

It is clear that a structure in which its components are held together entirely or mainly by covalent bonds will need a three-dimensional system of such bonds. From the tabulation above, it can be seen that sp^3 hybrid orbitals fulfil this requirement.

The best example of a covalent solid is the diamond form of carbon, shown in Figure 3.11. In this structure, each carbon atom is bonded to four other carbon atoms by sp^3 covalent hybrid bonds, thus providing an infinite, three-dimensional network of linked carbon atoms. Other elements, silicon, germanium and α-tin (grey), in periodic group 14 (previously, group IV) exhibit this structure type. Tin is dimorphous, β-tin (white) being metallic (Figure 3.12), as is lead (see also Section 4.5ff). In this periodic group, we can trace a continuous change in bond type from covalent to metallic. This change is paralleled by changes in physical properties, such as the electrical resistivity ρ, shown in Table 3.2. The varying bond types from which these results derive show again how it is

Table 3.2 Variation in resistivity in periodic group 14

	C (diamond)	Si	Ge	Sn (grey)	Sn (white)	Pb
$\rho/\Omega\,m$	5×10^{12}	2×10^{3}	0.5	1×10^{-5}	1×10^{-7}	2×10^{-7}

not possible to set up unique demarcations between the known types of compound.

Other solids that are predominantly covalent include silicon carbide SiC, which exists in several polymorphs, two of them exhibiting the wurtzite and blende structure types shown by zinc sulfide (see Figure 2.5). These two structures types may be visualized as a superposition of puckered sheets: in zinc blende the sheets are identical, although translated, whereas in wurtzite the successive layers are rotated by 180°. The sequence of layers are, thus, $A, A, A, A, \ldots$ and $A, B, A, B, \ldots$ respectively. The other forms of silicon carbide exhibit these same puckered sheets in other, more complex stacking modes, one repeating only after thirty-three layers.

The Si—O bond in silicon dioxide has been calculated to have at least 60% covalent character (67% from electronegativities – Section 3.3.1), so that the oxides of silicon may be classed as covalent solids. Figure 3.13a shows a structure for silica glass. It contains

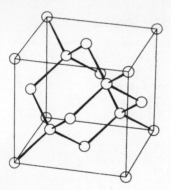 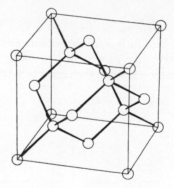

Figure 3.11. Stereoview of the unit cell and environs of the diamond allotrope of carbon. The space group is cubic, $Fd3m$, with eight atoms per unit cell. The carbon atoms occupy Wyckoff a positions: $(0, 0, 0; \frac{1}{4}, \frac{1}{4}, \frac{1}{4}) + F$. This structure is adopted also by silicon, germanium and gray tin.

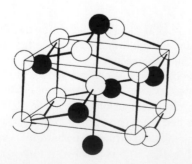 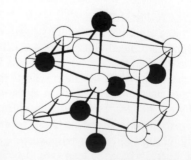

Fig. 3.12. Stereoview of the unit cell and environs of white tin (β-Sn). White tin is metallic, with four nearest neighbours and two at a slightly longer distance; these six atoms are shown in black on the diagram. The transition to a true metal in this periodic group is complete at lead.

the structural unit [SiO₄], as do all forms of silica, but there is no long-range order in the structure, and it is an amorphous solid. In contrast, we consider the structure of quartz.

There are two forms of quartz: α-quartz is hexagonal, space group $P3_121$, with silicon in Wyckoff a positions and oxygen in c. Quartz is optically active in the solid state, and in l-α-quartz the direction of travel of light through it follows the right-handed screw direction of the structure, Figure 3.13b.

If α-quartz is maintained at a temperature of ca 580 °C, it transforms to β-quartz. This compound is also hexagonal, but with space group $P6_222$; the silicon atoms are in Wyckoff c and oxygen in j positions of this space group. Figure 3.14 shows a stereoview of the structure of β-quartz. As in other forms of silicon dioxide the [SiO₄] tetrahedral units of structure are linked throughout by sharing a corner, that is, with an oxygen atom

Fig. 3.13. Arrangements of [SiO₄] structural units; the darker spheres represent the silicon atoms. (a) Silica glass; long-range order is absent. [Crown copyright. Reproduced from NPL Mathematics Report Ma62 by R J Bell and P Dean with the permission of the Director.] (b) Unit cell and environs of the α-quartz structure.

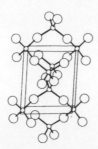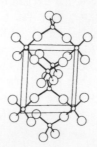

Figure 3.14. Stereoview of the unit cell and environs of the β-quartz structure; circles in order of decreasing size represent O and Si atoms. The space group is hexagonal, $P6_222$: the Si atoms are on Wyckoff positions c and the O atoms on j. In both forms of crystalline quartz, the screw nature of the symmetry (3_1 and 6_2) provide the structural conditions for optical activity in the solid state.

common to two tetrahedra.

Cristobalite and tridymite, each of which has two modifications, are other forms of silicon dioxide. They resemble closely the structures of zinc blende and wurtzite, respectively. If, in wurtzite, the zinc and sulfur atoms are both replaced by silicon, and oxygen atoms introduced mid-way between each pair of silicon atoms, the (idealized) structure of tridymite is obtained. In a similar manner, cristobalite may be imagined to be derived from zinc blende. A detailed examination has shown that the oxygen atoms are displaced slightly from the straight lines joining pairs of silicon atoms; this arrangement is an indication of the directional, covalent nature of the Si—O bond.

3.3.1 Ionic-covalent character in bonds
We introduced electronegativity in Section 2.2; the smaller the difference in the electronegativities of the two species forming a bond, the greater is its covalent character. The relationship between electronegativity difference $|\Delta\chi|$ and the fractional ionic character, or **ionicity**, q is given approximately, for q less than ca 0.5, by the equation

$$(3.48) \qquad q = 0.0695|\Delta\chi|^2 + 0.108|\Delta\chi| .$$

This equation was established from dipole moment data on the hydrogen halides. The dipole moment p is, strictly, a vector property, but for a linear species it is given by

$$(3.49) \qquad p = qed ,$$

where e is the electronic charge and d is the bond length, whence we derive Table 3.3 using (3.49)–(3.50) and known data, with q_{exprt} from (3.49). The fraction λ of the ionic contribution to the wavefunction of the hydrogen halides is given by

$$(3.50) \qquad \lambda^2 = q_{\text{exprt}}/(1 - q_{\text{exprt}}) .$$

In the case of sodium chloride, for example, $|\Delta\chi|$ is 2.23, which leads to a value of q of 0.58, which is much too low. Zinc oxide has a $|\Delta\chi|$ of 1.79, so that q is 0.41. In the

Table 3.3 Ionicity of the hydrogen halides

	HF	HCl	HBr	HI
$10^{30}\, p/\,\mathrm{C\ m}$	6.07	3.44	2.77	1.50
$10^{9}\, d/\mathrm{nm}$	0.0927	0.127	0.141	0.161
q_{exprt}	0.409	0.169	0.123	0.0582
q_{χ}	0.410	0.167	0.121	0.0633
λ	0.832	0.451	0.375	0.249

case of zinc sulfide q is 0.16, so that on this basis the zinc blende and wurtzite structure types may be termed mainly covalent. However, if we calculate the lattice energy for zinc blende with the electrostatic model $U(r_e)$ is approximately 3867 kJ mol^{-1}, whereas on the thermodynamic model it is 3709 kJ mol^{-1}; the discrepancy of *ca* 4% here is not high for this kind of calculation. Hence, the dilemma of classification will always obtain.

EXAMPLE 3.5. If the dipole moment of chlorobenzene is 1.69 D, what values would you expect for the dipole moments of the three dichloro-susbtituted benzenes? The answers are at the end of the chapter.

3.4 PROPERTIES ASSOCIATED WITH COVALENT SOLIDS
Covalent solids are characterized by a directional character in the bonding; we have seen how this feature arises in terms of the overlap of atomic orbitals and hybrid atomic orbitals. The number of nearest neighbours of any atom tends to be small, so that the structures have an 'open' character. Covalent solids are mechanically strong and hard, with high melting-points and low expansivities. They are electrical insulators in the solid and when molten, this latter property being in clear contrast to ionic compounds. Covalent solids, when transparent, are highly refracting to light. They are generally chemically unreactive and insoluble in all usual solvents.

ANSWERS TO EXAMPLES 3
3.1. $\mu_{\mathrm{HBr}} = (m_{\mathrm{H}} \times m_{\mathrm{Br}})/(m_{\mathrm{H}} + m_{\mathrm{Br}}) = 0.9953$. Notice that when $m_1 \ll m_2$, μ tends to m_1.

3.2. $\int_0^a A^2 \sin^2(n\pi x/a)\ \mathrm{d}x = 1 = A^2 \int_0^a [1 - \cos(2n\pi x/a)]/2\ \mathrm{d}x = A^2 a/2$, so that $A = (2/a)^{\frac{1}{2}}$.

3.3. Probability $P = 4\pi \int_{1.10a_0}^{1.11a_0} \psi^2 r^2\ \mathrm{d}r = 4\pi \int_{1.10a_0}^{1.11a_0} \exp(-2r/a_0)\ r^2\ \mathrm{d}r$. By the reduction formula,

$\int r^2 \exp(cr)\ \mathrm{d}r = (r^2/c) \exp(cr) - (2/c) \int r \exp(cr)\ \mathrm{d}r = (r^2/c) \exp(cr) - (2/c)[\int r \exp(cr)\ \mathrm{d}r] = (r^2/c)$
$\exp(cr) - (2/c)[(1/c) \int \exp(cr)\ \mathrm{d}r] = \exp(cr)[\ r^2/c - 2r/c^2 + 2/c^3]$, where $c = -2/a_0$. Evaluating between the limits $1.10a_0$ and $1.11a_0$, $P = 4\pi \times 4.2695 \times 10^{-4} = 3.565 \times 10^{-3}$. Note that very nearly the same result is obtained from the expression $4\pi r^2 \psi^2 \mathrm{d}r$, with $r = 1.10a_0$ and $\mathrm{d}r = 0.01a_0$. E(ground state) $= -2.1787$ J; $I = 1312.0$ kJ mol^{-1}.

3.4. $N_1 N_2 \int_0^\pi \sin(\theta)\ \cos(\theta)\ \mathrm{d}\theta = N_1 N_2 \int_0^\pi \sin(2\theta)/2 = 0$, so that the functions are orthogonal. For

normalization, $N_1^2 \int_0^\pi \sin^2(\theta)\ \mathrm{d}\theta = 1 = N_1^2 \int_0^\pi 2[\cos(\theta) - 1]/2\ \mathrm{d}\theta = \pi$, so that $N_1 = (1/\pi)^{\frac{1}{2}}$. N_2 has the same value.

3.5. Since the dipole moment is a vector property, the value for a dichlorobenzene molecule p_2 is given by $p_2 = [p_1^2 + p_1^2 + 2p_1^2 \cos(\theta)]^{\frac{1}{2}}$, where p_1 is the dipole moment of chlorobenzene and θ is the angle between the two C–Cl bonds in a dichlorobenzene. Thus, 1,2-dichlorobenzene, $p = 2.93$ D; 1,3-dichlorobenzene, $p = 1.69$ D; 1.4-dichlorobenzene, $p = 0$.

PROBLEMS 3

3.1. What is the uncertainty in (a) the speed of a cricket ball of mass 0.5 kg that is less than 1 μm from the bat, and (b) the speed of an electron that is 1 μm from a reflecting plane in a crystal?

3.2. The wavefunction for the $1s$ electron in the hydrogen atom may be written as $N \exp(-r/a_0)$, where N is the normalizing constant and a_0 is the Bohr radius for hydrogen. Determine N.

3.3. The molecule below may be treated as the 'box'

$$ -\overset{..}{N}-C=C-C=C-C=\overset{+}{N}- , $$

where the mean bond length is 0.14 nm. Calculate the value of the lowest-energy transition.

3.4. If the probability function for the $1s$ electron in the hydrogen atom is given by $4\pi r^2 (1/\pi a_0^3)^{-\frac{1}{2}} \exp(-2r/a_0)$, find the position of the electron for maximum probability.

3.5. What are the electron configurations of the species (a) C, (b) O, and (c) F?

3.6. What are the symmetry classification of the bonding molecular orbitals in (a) HF, (b) F^2, and (c) CO?

3.7. What bond angle is expected between hybrid orbitals that are of the form $(1/\sqrt{3})s + \sqrt{(2/3)}p$?

3.8. The dipole moment of the water molecule is 1.8 D. What are the nominal charges of the hydrogen and oxygen atoms, given O—H = 0.96 and H–O–H = 104.4°?

3.9. The cubic unit cell of the diamond structure has a side a of 0.3567 nm. Calculate the C—C bond distance in this species.

3.10. Low temperature, α-quartz is hexagonal, with $a = 0.4913$ nm and $c = 0.5405$ nm. The atoms have the following positions (origin shift to 0, 0, 1/3):

3 Si   0.465, 0, 0; 0, 0.465, 2/3; −0.465, −0.465, 1/3.

6 O   0.415, 0.272, 0.120; −0.143, −0.415, 0.453; −0.272, 0.143, 0.787;
           0.143, −0.272, −0.120; 0.272, 0.415, 0.547; −0.415, −0.143, 0.213.

Determine (a) the Si—O bond distance and the *two* different, nearest-neighbour O—O distances. It may help to make a careful sketch of the structure on the x,y plane. List the coordinates of the atoms involved in the distances calculated.

CHECKLIST 3

At the end of this chapter, you should be able to:

1. Write and explain the 1- and 3-dimensional Schrödinger equation for an electron;
2. Explain the uncertainty principle;
3. Understand and calculate the reduced mass of a species;
4. Describe Born's interpretation of the wavefunction ψ, and normalization;
5. Understand the solution of the electron-in-a-box and how it leads to quantization;
6. Appreciate meaning of zero-point energy and its relation to dissociation energy;
7. Define radial and angular parts of the one-electron wavefunction (atomic orbital);
8. Calculate the positional probability of an electron in the hydrogen atom;
9. Understand the terms shell, sub-shell and their notations;

10. Describe the Pauli exclusion principle, the Aufbau principle and Hund's rule;
11. Understand the Born-Oppenheimer approximation for wavefunctions, and the variation method for multi-electron atoms;
12. Define the LCAO methods for molecular wavefunctions;
13. Solve, in principle, a set of secular equations;
14. Define orthogonality, overlap integral, Coulomb integral and resonance integral;
15. Apply the variation method to a simple species, e.g. H_2^+;
16. Understand bonding and antibonding molecular orbitals, and electron density;
17. Classify molecules according to their symmetry;
18. Define hybrid orbitals, and explain with reference to H_2O and CH_4;
19. Describe simple covalent solids, such as diamond and other elements of group 14, zinc blende and wurtzite, and polymorphs of silica;
20. Calculate the ionic character of covalent bonds, in relation to electronegativity and dipole moment;
21. Describe the structural and physical characteristics associated with the covalent bond.

4

Looking at metal structures

4.1 INTRODUCTION

Approximately three-quarters of the known elements are metals, yet their structure types are few and, geometrically, quite uncomplicated. On the one hand, we can think of the structures of metals in terms of the packing of identical atoms, assumed to be effectively spherical. This approach will enable us to discuss their structure types, and some of their properties in a qualitative manner. On the other hand, and in more detail, we will find that the metallic bond may be regarded as an extreme example of conjugation, where the bonding electrons in molecular orbitals of the the metal range throughout the whole structure. A simple theoretical approach, the free-electron theory, is useful in explaining transport properties of metals.

4.2 FREE-ELECTRON (CLASSICAL) THEORY OF METALS

The free-electron theory divides the electrons in a metal into two groups: there are those forming a **core** around a lattice of positive ions, and those that are free to move, the **conduction** (or **valence**) electrons. This theory gives rise to the picture of a lattice of positive ions with their core electrons in closed shells, surrounded by an **electron gas**. Cohesion arises from the interaction of the electron gas with the lattice of cations, and this model can throw light on the properties of thermal conductivity, electrical conductivity and opacity to light and longer-wave radiations.

4.2.1 Classical solids

The heat capacity at constant volume C_v of a monatomic solid is approximately 25 J K^{-1} mol^{-1}, the value given by the Dulong and Petit law. Figure 4.1 shows the general variation of constant volume heat capacity with temperature. If we assume that the atoms in a crystalline metal vibrate under simple harmonic motion with a frequency ν, then the total energy E, the sum of the kinetic energy and potential energy, may be given in the form

$$(4.1) \qquad E = \tfrac{1}{2}mv^2 + 2\pi^2\nu^2mx^2,$$

where m is the mass of the atom, and v and x are, respectively, the instantaneous speed and linear displacement from equilibrium. The mean vibrational, thermal energy for a classical oscillator at a temperature[7] T is k_BT, where k_B is the Boltzmann constant. For L such oscillators, where L is the Avogadro constant, and generalizing to three dimensions, the mean total molar energy becomes

$$(4.2) \qquad \overline{E_m} = 3Lk_BT = 3RT.$$

Since the constant volume heat capacity is defined by $(\partial E/\partial T)_v$, it follows that the constant volume heat capacity is given by

$$(4.3) \qquad C_v = 3R,$$

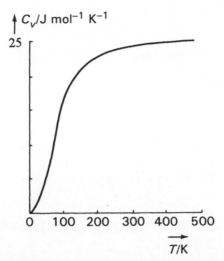

Fig. 4.1. Variation of the constant volume heat capacity with temperature for a monatomic solid.

Table 4.1 Debye temperatures for some elements

Element	Θ/K
C (diamond)	2230
C (graphite)	420
Ge	370
Na	158
K	91
Al	428
Cu	343
Ag	225
Pb	105

which evaluates to approximately 25 J K^{-1} mol^{-1}. This result applies well at temperatures greater than *ca* 250 K, but fails significantly at temperatures below *ca* 200 K. Refinements of the classical theory of heat capacity first by Einstein and then by Debye led to a revised equation which, for low temperatures, takes the form:

(4.4) C_v/J K^{-1} mol$^{-1} = 1944(T/\Theta)^3$,

where Θ is the Debye temperature; values of Θ for some common substances are given in Table 4.1.

EXAMPLE 4.1. What is the molar heat capacity at constant volume of aluminium at 10 K (a) under the Dulong and Petit law, and (b) under the Debye law? The answers are at the end of the chapter.

A mole of conduction electrons in a metal, behaving as free particles, would contribute a kinetic energy of $3RT/2$, or $1.5R$ to the constant volume heat capacity. The total value, including the vibrational contribution, would now be $4.5R$. For $\Theta/T > 0.7$, the deviation from (4.3) is no more than 10%. Since the contribution to the heat capacity from vibration is based on a satisfactory theory, it follows that the contribution from the electrons must be very small, a result at variance with the classical theory, and often termed the **heat capacity paradox**: we need a more detailed theory.

4.3 WAVE-MECHANICAL THEORY OF METALS

At first, we may treat the conduction electrons as free, but bound collectively to the positive ions with which they interact over the whole crystal. This assumption allows us to treat a conduction electron in a metal by the three-dimensional analogue of the electron-in-a-box (Section 3.2.1), but with a periodic potential energy so as to conform to an array of atoms having. the translations of a Bravais lattice.

With the electron at zero potential energy in a cubic box of side a, the wavefunction is the standing wave

$$(4.5) \qquad \psi(x,y,z) = (8/a)^{\frac{1}{2}} \sin(n_x\pi x/a) \sin(n_y\pi y/a) \sin(n_z\pi z/a) ,$$

where n_x, n_y and n_z are positive integers. Since the wave equation has to satisfy the periodic boundary conditions

$$(4.6) \qquad \psi(x,y,z) = \psi(x + a,y,z) = \psi(x,y + a,z) = \psi(x,y,z + a) ,$$

a suitable solution is the plane travelling wave

$$(4.7) \qquad \psi_k(r) = \exp(i\, k.r) ,$$

where, since the unit vectors i, j and k have equal magnitudes for a cubic box,

$$(4.8) \qquad r = i.(x + y + z) ,$$

and the **wave vector k** has components such that

$$(4.9) \qquad k = i.(k_x + k_y + k_z) ,$$

such that

$$(4.10) \qquad k_x = 0 \text{ or } \pm 2n_x\pi/a ;$$

similarly for the directions y and z. The components of k are the quantum numbers for the electron which, with the spin quantum number, provide the total of four quantum numbers needed to specify an electron under the Pauli exclusion principle. By the usual manipulation (see Section 3.2.1),

$$(4.11) \qquad E_k = [\hbar^2/(2m_e)](k_x^2 + k_y^2 + k_z^2) .$$

The energy levels given by (4.11) approximate to a **band**, because they are very close. At 0 K, the electrons occupy energy levels up to a value known as the **Fermi energy** E_F. The Fermi energy per atom is of the order of 10^{-18} J, and if we assume that the mean kinetic energy is $0.6E_F$ (see Section 4.3.3), then the corresponding Fermi temperature T_F would be approximately 15×10^3 K. This result means that, in all normal circumstances, the heat supplied to a crystal will lead to a temperature so much less than T_F that it will

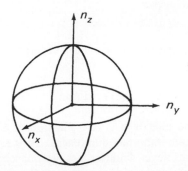

Fig. 4.2. Sphere of radius k_F in (reciprocal) k-space, on the directions of n_x, n_y and n_z as axes. The axes are mutually perpendicular, and the triplet (n_x, n_y, n_z) identifies a lattice point (energy state) in k-space (see Appendix 3).

not perturb the electron distribution; then, the electron contribution to the heat capacity becomes vanishingly small, as required from our discussion on the heat capacity paradox.

The magnitude of the wave vector is related to the wavelength by

(4.12) $$k = 2\pi/\lambda .$$

By using the de Broglie equation relating momentum to wavelength, $p = h/\lambda$, (4.11) may be transformed to

(4.13) $$E_k = p^2/2m_e ,$$

which shows a parabolic relationship between energy and momentum. The same result is given by the simpler free-electron theory.

We consider adding electrons to a lattice of metal ions. For each energy state governed by n_x, n_y and n_z, an electron with spin $+\frac{1}{2}$ or $-\frac{1}{2}$ can be accommodated. As $|k|$ increases so does the energy associated with the state represented by n_x, n_y and n_z. In the ground state system of N electrons, the occupied states may be represented by the points of a lattice in k-space (the space of n_x, n_y and n_z) lying within a sphere of radius k_F, the Fermi sphere shown in Figure 4.2. The energy at the surface of this sphere represents the Fermi energy E_F, given by

(4.14) $$E_F = \hbar^2 k_F^2/(2m_e) .$$

The term $2\pi/a$ defines the side of the primitive unit cell of the (reciprocal) lattice in k-space which may be identified with a single energy state. Since *one* lattice point is associated with the volume of any P unit cell, the number N of states in a sphere of radius k_F is

(4.15) $$N = 2(4\pi k_F^3/3)/(2\pi/a)^3 .$$

The factor 2 arises because each lattice point (energy state) can accommodate two electrons of spins $\pm\frac{1}{2}$.

EXAMPLE 4.2. Determine the number of conduction electrons per unit volume (N/V) in real space for the face-centred unit cell of copper, $a = 0.361$ nm, and the corresponding value of E_F. The answers are at the end of the chapter.

4.3.1 Density of states

The density of states $g(E)$ refers to the number of energy states of the solid lying in the range E to $E + dE$, that is,

$$(4.16) \qquad g(E) = dN/dE,$$

and Figure 4.3 shows the plot of $g(E)$ as a function of E. It is parabolic, as found in (4.13), but now with certain important differences. At 0 K, energy states are filled to the cut-off at the Fermi energy level; higher states are unoccupied. At higher temperatures, the thermal energy causes some electrons to rise above the Fermi level. Only those electrons with an energy of approximately $k_B T$ less than E_F are affected in this way. Most electrons are unavailable for thermal excitation at low temperatures, because of the stability of paired electrons in filled energy states.

The constant volume heat capacity of a metal at temperatures less than *ca* 15 K may be written as the sum of two terms, one connected with atomic (lattice) vibrations, and the other with the electrons. Thus, we have

$$(4.17) \qquad C_v = aT^3 + bT,$$

where a and b are constants. From (4.4), we know that C_v/T is a linear function of T, and the plot will have a slope of a and an intercept of b at $T = 0$ K. Experiments with metallic copper gave a value of *ca* 7×10^{-4} J K^{-2} mol^{-1} for b. Thus, its contribution to C_v at 25 °C is $7 \times 10^{-4} \times 298$, or 0.21 J K^{-1} mol^{-1}, approximately 0.8% of the classical Dulong and Petit value of *ca* 25 J K^{-1} mol^{-1}.

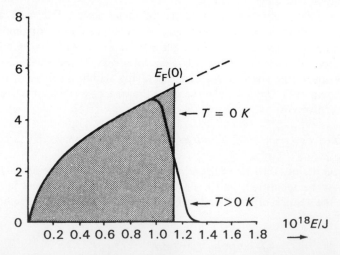

Fig. 4.3. Density of states function for copper; the sharp cut-off occurs at 0 K. At higher temperatures, some electrons are promoted above E_K, so that the curve becomes rounded, with a 'tail' at higher values of E.

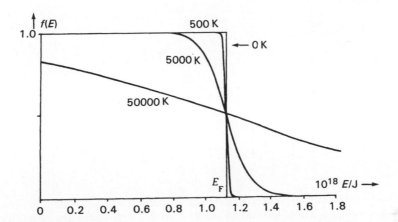

Fig. 4.4. Fermi-Dirac distribution function $f(E)$ for copper; the curves for different temperatures intersect at $f(E) = \frac{1}{2}$, when $E = E_F$.

4.3.2 Fermi-Dirac distribution
The energies of electrons in metals follow a Fermi-Dirac distribution:

(4.18) $$f(E) = [\exp(E - E_F)/(k_B T) + 1]^{-1}.$$

Thus, the number of states of energies lying between E and $E + dE$ is given by

(4.19) $$N(E) = g(E) f(E) \, dE,$$

where $g(E)$ is the density of states function (4.16).

The form of (4.18) is shown by Figure 4.4. At 0 K, $f(E)$ is unity for $0 \le E \le E_F$, and changes discontinuously to zero for $E > E_F$. As the temperature increases, the curve becomes progressively rounded near the value of E_F. The function $N(E)$ at 0 K may be considered as $g(E)$ multiplied by a function that is unity for $0 \le E \le E_F$, and zero otherwise, and corresponds to the shaded portion of Figure 4.3. At temperatures greater than 0 K, $f(E)$ is less than unity for $E < E_F$, and greater than zero for $E > E_F$; for $E = E_F$, $f(E) = \frac{1}{2}$ at all temperatures. The high-energy regions of Figure 4.4 for $T > 0$ correspond to $(E - E_F) \gg (k_B T)$, or $(E - E_F)/(k_B T) \gg 1$; then,

(4.20) $$f(E) = \exp[-(E - E_F)/(k_B T)],$$

which corresponds to the classical, Maxwell-Boltzmann distribution applicable to the free electron gas discussed in Section 4.2.

4.3.3 Mean electronic energy
We are now in a position to calculate the mean energy of the electrons in a metal. From the standard formula, we have

(4.21) $$\overline{E_{el}} = \frac{\displaystyle\int_0^\infty E \, g(E) \, f(E) \, dE}{\displaystyle\int_0^\infty g(E) \, f(E) \, dE}.$$

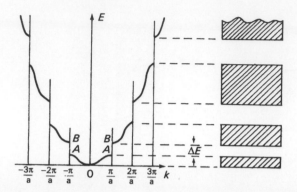

Fig. 4.5. Quadratic relationship between E_k and k but with gaps at values of k for which the Bragg equation holds, leading to a band structure of energy states. The first energy gap AB sets the value of ΔE between the first two energy bands.

At 0 K, $f(E)$ is unity from 0 to E_F but zero otherwise; since $g(E)$ is directly proportional to E, (4.21) takes on a simple expression and leads to the value of $3E_F/5$ for the average energy of the electrons:

(4.22)
$$\overline{E_{el}} = \frac{\displaystyle\int_0^{E_F} E^{3/2}\, dE}{\displaystyle\int_0^{E_F} E^{1/2}\, dE}.$$

By simple integration, it is straightforward to show that this expression evaluates to $3E_F/5$, the value we used for $\overline{E_{el}}$ in Section 4.3.

4.4 BAND THEORY

Neither of the free-electron theories that we have discussed so far explains satisfactorily the large variations in electrical resistivity found among metals, semiconductors and insulators. Band theory uses the Schrödinger equation with a periodic potential, and leads to a wave (Bloch) equation of the form

(4.23)
$$\psi_k(x) = E_k(x)\,\exp(ikx) ,$$

where $E_k(x)$ is an energy function of periodicity a along the x direction; similar expressions hold for the directions y and z, with the periodicities b and c, respectively. The dependence of energy on k is still quadratic, but discontinuities in the energy arise for $k = \pm n\pi/a$ that lead to a band structure. At most values of k the electrons behave as though they were free, but when $k = \pm n\pi/a$ the condition for Bragg reflection, in a periodicity a, of electron waves at an angle θ occurs, namely

(4.24)
$$2a \sin(\theta) = n\lambda ,$$

with $\sin(\theta) = 1$ and $k = 2\pi/\lambda$. The first-order reflection ($n = 1$) corresponds to $k = \pm\pi/a$, and the region between $+\pi/a$ and $-\pi/a$ defines the **first Brillouin zone**. Figure 4.5 shows

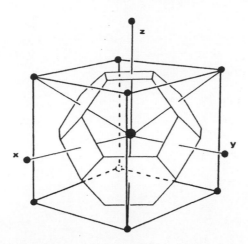

Fig. 4.6. Wigner-Seitz (cubo-octahedron) cell from a cubic I lattice unit cell. In (reciprocal) k-space, the same polyhedron forms the first Brillouin zone. No other bisecting planes can be drawn that would intersect the polyhedron.

a plot of energy against k, with its discontinuities leading to a band structure of closely-spaced energy levels. The energy gap ΔE between bands is intimately connected with the electrical conductivities of metals, and other solids.

Τhe concept of Brillouin zones can be extended to three dimensions. The zone boundaries are determined by the regions in k-space where the Bragg equation (for diffraction) is satisfied, so that they are governed by crystal structure type. In three dimensions, the first Brillouin zone corresponds to the smallest volume in k-space enclosed by planes that bisect normally the shortest vectors drawn from the origin of k-space to points n_x, n_y, n_z.

Consider a body-centred cubic lattice unit cell in Bravais space, Figure 4.6. We take the lattice points at the eight corners and those of the body-centring points in the six adjacent unit cells of the lattice, and draw the perpendicularly bisecting planes. The enclosed figure is a combination of the cube and octahedron polyhedra, a Wigner-Seitz cell[20-23], also called the cubo-octahedron. If we carry out the same construction in k-space then, because the reciprocal of an I unit cell is an F unit cell[4], the construction leads to the same shaped figure which now constitutes a first Brillouin zone, a Wigner-Seitz cell in reciprocal space (see also Appendix 3).

4.4.1 Molecular orbitals and energy bands
In a metal, the number of atoms in a small sample crystal is very large, ca 10^{15}, and the electrons are delocalized, or free to move, around the whole crystal. The molecular orbitals are now **conduction orbitals**, and metallic character depends upon their overlap.

Consider a row of N atoms, each containing a single conduction electron. An atom with a single s electron forms a single s orbital. When a second atom is brought into overlap with it, they form a bonding and an antibonding molecular orbital. A third atom overlaps the second strongly and the first less strongly to form three molecular orbitals.

This process can be envisaged to continue for all N atoms, so that a total of N molecular orbitals is obtained. When N is large the energy gap between successive energy states is so small that they form effectively a band. The energy of the sequence of N atoms can be approximated by the Hückel molecular-orbital method[16], with the result

$$(4.25) \qquad E_n = \alpha + 2\beta\ \cos[n\pi/(N+1)]\ ,$$

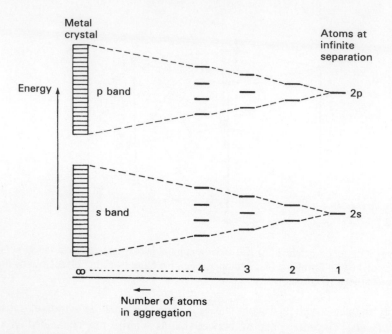

Fig. 4.7. Schematic representation of the formation of *s* and *p* bands in a metal crystal. The lowest and highest levels in each band correspond to completely bonding and completely antibonding situations, respectively.

where *n* ranges from 1 to *N*. It is evident that $(E_{n+1} - E_n)$ becomes vanishingly small as *N* tends to a large number.

A band formed from *s* orbitals is termed an *s* band, and that from *p* orbitals a *p* band. Normally, the *p* band lies above the *s* band in energy, the gap between them being the value of ΔE for those two bands; Figure 4.7 illustrates this molecular-orbital picture of a metal.

4.5 METAL STRUCTURES

The extreme delocalization of the valence electrons in metals effectively removes the directional character of the orbitals that we associated with the molecular-orbital description of the covalent bond. Hence, the structures of metals are determined principally by space-filling criteria. The metals of periodic groups 1, 2, 11, the transition-type metals and certain others conform to one or other of three simple arrangements: the close-packed *F* cubic (*A*1), the close-packed *P* hexagonal (*A*3), and the less closely packed *I* cubic structures.

The two close-packed structure types, *A*1 and *A*3, represent the two modes of regular, closest packing of identical spheres. Figures 4.8 and 4.9 illustrate these two structure types. In both *A*1 and *A*3, a first layer is obtained by placing spheres in closest contact, such that their centres form the apices of equilateral triangles. A second layer, identical with the first, is added such that the spheres of that layer rest in the depressions of the first layer. A third, exactly similar layer can now be added in one of two ways. On the one hand, if it is arranged so that the spheres in the third layer lie over 'holes' in both the first and second layers, then the close-packed cubic structure *A*1 is obtained. On the other hand, if it is positioned such that the spheres in the third layer lie exactly above those in the first layer, the close-packed hexagonal structure *A*3 is formed. The sequence of layers in the two arrangements is, thus:

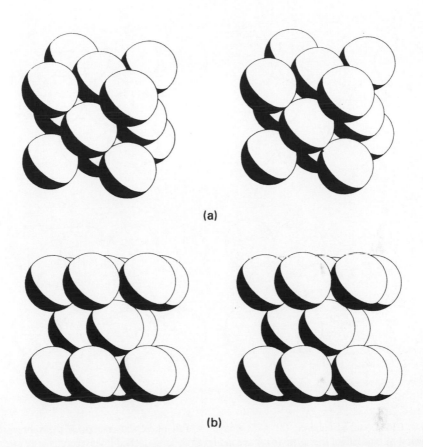

(a)

(b)

Fig. 4.8. Stereoviews of the unit cells and environs of closest packed identical spheres. (a) Close-packed cubic, $A1$; space group $Fm3m$, with atoms in Wyckoff a positions of symmetry $m3m$. (b) Close-packed hexagonal, $A3$; space group $P6_3/mmc$, with atoms in Wyckoff c positions of symmetry $\bar{6}m2$. (By change of origin, the atomic coordinates become 0, 0, 0 and 1/3, 2/3, 1/2.) The coordination number is 12 and the packing fraction is 0.74 in each structure.

$$A1: A\,B\,C\,A\,B\,C\,A \,...$$
$$A3: A\,B\,A\,B\,A\,B\,A \,...$$

In both of these structures, the coordination is twelve-fold, the maximum number for *regular* packing of identical spheres. From Figure 4.8a, the spheres are in diametrical contact along a face diagonal of the cube. If the side of the cube is a, the volume occupied per sphere is $a^3/4$, because there are four atoms per F unit cell. The volume of each sphere is $4\pi r^3/3$, where r is the radius of the sphere. Hence, the fraction of space fully occupied, the **packing fraction**, is $(4\pi r^3/3)/(8r^3/\sqrt{2})$, or 0.74. The same value may be found for the $A3$ structure.

In the body-centred cubic $A2$ structure, the coordination number is eight. This structure is less closely packed than the CPC and CPH structures. From Figure 4.10, we can see that the body-diagonal contact means that $a\sqrt{3} = 4r$, so that the packing fraction is now 0.68.

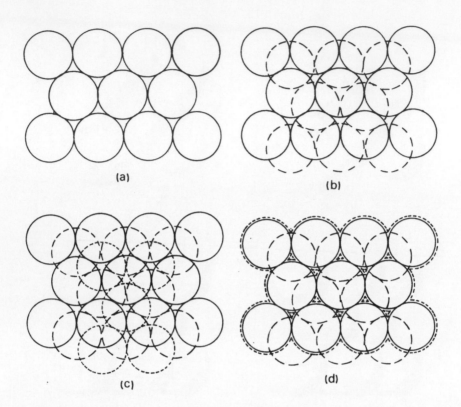

(a)

(b)

(c)

(d)

Fig. 4.9. Close packed spheres. (a) A first layer *A*, full lines; the centres of the spheres form a succession of equilateral triangles. (b) A second layer *B*; long-dashed lines; there is only *one way* in which the second layer can be close-packed on to the first. (c) A third layer *C*, short-dashed lines; the fourth layer is then placed exactly above the first layer *A*, so that the sequence is *ABCA...* for the CPC, or *A*1, structure. (d) A third layer, short-dashed lines (the spheres are shown with slightly larger radius, for clarity); now, the third layer lies exactly above the first layer *A*, leading to the CPH, or *A*3, structure.

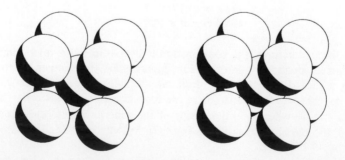

Fig. 4.10. The body-centred cubic *A*2 stucture, space group *Im3m*, with atoms in Wyckoff *a* positions of point-group symmetry *m3m*. The coordination number is 8, and the packing fraction for this strtucture type is 0.68.

Table 4.2 Radii of metals/nm, and their structure types

Li	0.152	$A1, A3$	Sr	0.215	$A1, A3$
Na	0.186	$A3$	Ba	0.222	$A2$
K	0.227	$A2$	Cu	0.128	$A1$
Rb	0.248	$A2$	Ag	0.144	$A1$
Cs	0.265	$A2$	Au	0.144	$A1$
Be	0.112	$A3$	Fe	0.124	$A1, A2$
Mg	0.160	$A3$	Co	0.125	$A1, A3$
Ca	0.197	$A2, A3$	Ni	0.125	$A1, A3$

Table 4.3 Dependence of metal radius on coordination number

Coordination number	Relative radius
12	1.00
8	0.97
6	0.96
4	0.88

4.5.1 Metal radii

From Figures 4.8 and 4.10, it is evident that if we have a knowledge of the unit cell dimensions (from X-ray analyses), we can calculate the radii of metals. Thus, copper, which is a CPC structure has the dimension a of 0.3615 nm. We have used the relation for this structure type $a\sqrt{2} = 4r$, so that $r_{Cu} = 0.128$ nm. Table 4.2 list a selection of metal radii, as found from the unit cell dimensions of their structures, listed for twelve-fold coordination. Certain metals exist in polymorphic modifications and, not surprisingly, we find a dependence of radius on coordination number, as was the case for the radii of ions. Table 4.3 lists the relative radii, taking twelve-fold coordination as the standard.

4.5.2 Interstices in metal structures

In the $A1$ and $A3$ close-packed structures, each sphere has tetrahedral **interstitial sites**, or 'holes' as its (four) nearest neighbours and octahedral interstices as its (six) next-nearest neighbours. The radius of an interstitial hole is less than that of the spheres forming it, and the radius ratio limits of 0.225 and 0.414 apply, respectively, to the four-fold and six-fold coordination sites. In the close-packed hexagonal $A3$ structure, the interstices lies directly above one another, forming rows parallel to the six-fold (z) symmetry axis. It is a useful aid to the study of metal structures and their interstitial site to make up three or four layers of close-packed spheres, using polystyrene or table-tennis balls to build up the $A1$ and $A3$ structure types.

EXAMPLE 4.3. Determine the shortest distance between the sites for a tetrahedral and an octahedral hole in a cube of side a. The answer is at the end of the chapter.

4.6 ALLOYS

Alloys of metals have been known from very early times. They differ in their properties from other chemical compounds. They are generally prepared by simply melting the constituents together in the correct proportions, and they tend to retain many of the properties of the metals from which they are formed. The number of alloys is vast and

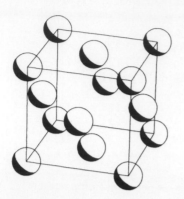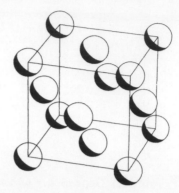

Fig. 4.11. Stereoview of the unit cell and environs of the random solid solution of copper in gold. Each sphere represents, effectively, a fraction x of copper and $(1 - x)$ of gold, so that the symmetry, statistically, corresponds to space group $Fm3m$.

Table 4.4 Classification of binary alloys

T									B_1		B_2		
										Al	Si		S
Sc	Ti	V	Cr	Mn	Fe	Co	Ni	Cu	Zn	Ga	Ge	As	Se
Y	Zr	Nb	Mo	Tc	Ru	Rh	Pd	Ag	Cd	In	Sn	Sb	Te
La	Hf	Ta	W	Re	Os	Ir	Pt	Au	Hg	Tl	Pb	Si	Po

their compositions often variable, so it is helpful to formulate a system for their classification.

We have discussed the band theory in application to pure metals. In the presence of a second component, the translational symmetry may be perturbed. This effect is small if the two components are from the same group of the periodic table, so that they have the same number of valence electrons, and also if they do not differ appreciably in their metallic radii.

Many important alloy systems involve that portion of the periodic table reproduced in Table 4.4. We list the elements of groups 3 to 11 as set T (true metals) and those of groups 12 to 16 as set B; it is convenient to sub-divide set B as shown in Table 4.4. Thus, we can envisage a possible four classes of alloy: T and T, T and B_1, T and B_2, B_1 and B_2. As we move through these classes, the alloys become progressively less metallic in character, and begin to show some similarity to ordinary chemical compounds. We shall consider binary alloys from some of these classes.

4.6.1 Copper—gold

Copper and gold form a T—T system; both have the close-packed cubic $A1$ structure type, with metallic radii 0.128 nm and 0.144 nm, respectively. They form a complete range of **substitutional solid solutions**, that is, where one component takes the place of the other at random throughout the structure. It has been found that provided the radii of two components do not differ by more than approximately 15%, complete solid solution can be expected. If the difference in radii is greater than this value the formation of solid

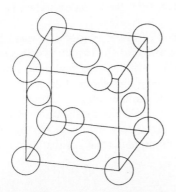 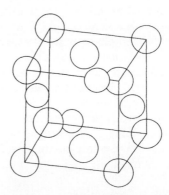

Fig. 4.12. Stereoview of the unit cell and environs of the ordered, tetragonal (pseudo-face-centred) structure of CuAu; the circles in order of decreasing size represent Au and Cu. The space group is *P4/m mm*, with Au in Wyckoff *a* and *c* positions, symmetry *4/m mm*, and Cu in *e*, symmetry *mmm*.

solution may be restricted or even inhibited.

If copper is added to pure gold, the molten mixture allowed to attain homogeneity and then quenched by sudden cooling, the resulting alloy has a face-centred cubic structure in which copper atoms have replaced gold atoms in a truly random manner. The overall symmetry remains *Fm3m*, although any individual unit cell may have a symmetry that departs from it. The unit cell dimension *a* follows Vegard's law, that is,

$$(4.26) \qquad\qquad a = Kc_{Cu},$$

where c_{Cu} is the concentration of added copper and K is a constant for the structure.

If, instead, the cooled samples are annealed, the same random replacement occurs until the composition reaches that corresponding to CuAu. Then, the atoms segregate into layers, forming a tetragonal structure which has an axial ratio *c/a* of 0.93, Figure 4.12. Continued replacement of gold by copper leads to random solid solutions of the tetragonal structure, and this continues until the composition is Cu_3Au. At that composition, a pseudo-face-centred cubic structure is obtained by annealing, Figure 4.13.

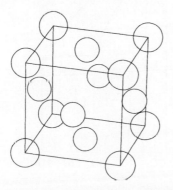 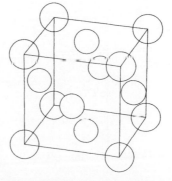

Fig. 4.13. Stereoview of the unit cell and environs of the ordered, cubic (pseudo-face-centred) structure of Cu_3Au; the circles in order of decreasing size represent Au and Cu. The space group is *Pm3m*, with Au in Wyckoff *a* positions, symmetry *m3m*, and Cu in *e*, symmetry *4/m mm*.

These ordered phases are termed **superstructures** (or **superlattices**). X-ray photographs show diffraction effects (**superlattice lines**), over and above those shown by the solid solutions of the same compositions.

An ordered phase must correspond to a system of lower energy than that of the solid solution of the same composition, although the entropy of the system is decreased. If the two components have very similar radii, substitution will involve negligible strain, so that the disordered system, of higher entropy, is preferred. This situation is observed for solid solutions of silver in gold. The contraction in atomic volume along the series of lanthanon metals (*lanthanide contraction*) results in gold having the same metallic radius as silver. There is, thus, no strain introduced into the structure on mixing silver with gold, so that the entropy term in the free energy function determines that the random solid solution is obtained.

4.6.2 Silver—cadmium

The silver—cadmium system is an example of a T—B_1 alloy; it is more complex than the copper—gold system. Although the radii are quite similar ($r_{Ag} = 0.144$ nm, $r_{Cd} = 0.152$ nm), the elemental structures and numbers of valence electrons are different; Figure 4.14 is the phase diagram for this binary system.

The α-phase represents pure silver, and it can take up to 42% of cadmium into solid solution. The structure is random face-centred cubic, with the unit cell side a proportional to the cadmium concentration. A β-phase then appears which becomes homogeneous at 50% cadmium; it is a random body-centred cubic structure

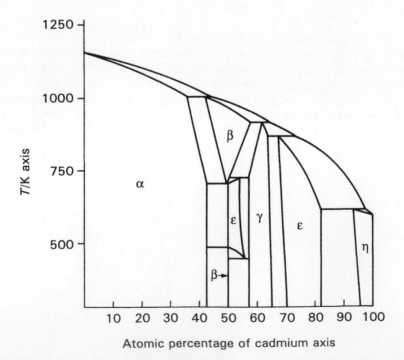

Fig. 4.14. Equlibrium phase diagram for the silver–cadmium alloy system, showing the ranges for the α (pure Ag), β, γ, ε and η (pure Cd) phases.

Beyond 50% Cd, a γ-phase develops, homogeneous at 57% Cd, which corresponds to the composition Ag_5Cd_8. This structure is complex cubic, with fifty-two atoms per unit cell, and is hard and brittle. At a cadmium concentration greater than 65%, the ε-phase develops. It is an approximately close-packed hexagonal arrangement, the lattice sites being occupied by cadmium and silver atoms in a random manner. The departure from the true $A3$ structure type appears in the c/a ratio of approximately 1.57.

Pure cadmium appears as the η-phase: it approximates to the ideal close-packed hexagonal structure ($c/a = 1.633$) with a c/a ratio of 1.9. It takes up only about 4% of silver into solid solution before the formation of the closely similar η-phase. Any strain that is present in a Cd—Ag solid solution is reduced more readily at the high cadmium region of the system than at high silver compositions. The compositions found have a theoretical basis, given originally in terms of the Hume-Rothery rules (see Section 4.6.4).

4.6.3 Nickel Arsenide

A T—B_2 alloy type of common occurrence is that of nickel arsenide NiAs. This structure exhibits space group $P6_3mc$ (see Figure 1.43), with two formula-entities per unit cell; the nickel atoms occupy Wyckoff a and the arsenic atoms the b sites. Figure 4.15 is a stereoview of the nickel arsenide structure. The coordination around nickel consists of an octahedral arrangement of arsenic atoms with a Ni—As bond length of 0.218 nm, and two other nickel atoms distant 0.250 nm ranged along the z axis. Although this structure is of the simple type MX, the atom sites are not interchangeable: the coordination of nickel around arsenic is that of a trigonal biprism. In nickel arsenide the c/a ratio is 1.39; had it been close packed, it would have been 1.63.

This structure is exhibited particularly by compounds of transition metals with elements of groups 15 and 16.

4.6.4 Binary compounds of B_1 and B_2 elements

Binary compounds involving elements from the sets B_1 and B_2 present certain difficulties in their classification. Significant solid solution formation is found only among B_1—B_1 compounds. The range of solid solution is greater in the Cd—Hg system than in that for the Cd—Zn system, because the metallic radii of cadmium and mercury are more similar than are those for cadmium and zinc. When both elements in the binary compound are of type B_2, the sodium chloride structure type (see Section 2.2ff) obtains, as with SnSb and PbSe, for example.

The $B1$—$B2$ compounds may be grouped according to Table 4.5. Those compounds on the left-hand side of the table tend to be more ionic in bond type, whereas those on the

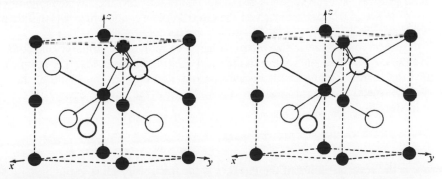

Fig. 4.15. Stereoview of the unit cell and environs of the structure of nickel arsenide NiAs; the larger circles represent As.

Table 4.5 Structure types of some B_1—B_2 binary compounds

Zinc blende						

Zinc blende							
BeS	ZnS	CdS	HgS	AlAs	GaAs	InAs
BeSe	ZnSe	CdSe	HgSe	AlSb	GaSb	InSb
BeTe	ZnTe	CdTe	HgTe			

Wurtzite	
ZnS	CdS
ZnSe	CdSe
ZnTe	-

right-hand side of the Table 4.3 are more covalent in character. From this point of view, compounds such as InAs and AlSb could have been included in Section 3.3, as covalent solids. However, in the present context, their relationship to chemically similar structures is better presented.

4.6.5 Hume-Rothery's rules for alloy systems

In a metal, the atoms are bonded somewhat indirectly by interaction with the delocalized valence electrons, so that the metal is almost unaffected by the replacement of a certain fraction of its atoms by those of another metallic species. The enthalpy of formation of an alloy AB, $\Delta H_f(AB)$, may be compared with the cohesive energies of its components. Thus, for an alloy composed of a fraction c of species A and $(1 - c)$ of B, this enthalpy of formation can be written as the difference between the cohesive energies of the alloy and its components (the pV terms cancel):

$$(4.27) \qquad \Delta H_f(AB) = U(AB) - [cU(A) + (1 - c)U(B)] .$$

Typically, $\Delta H_f(AB)$ is less than $0.1U(AB)$. If $\Delta H_f(AB)$ tends to zero, the (positive) entropy of mixing leads to solid solution formation, provided the two metals have the same crystal structure. In practice, a zero value for $\Delta H_f(AB)$ dos not actually occur, and the small value for this parameter manifests itself by partial solubility, the appearance of ordered phases, and intermetallic compound formation.

These features has been systematized in the Hume-Rothery rules, which embody three factors, atom size, electronegativity and the electron/atom ratio, which we shall consider briefly.

With respect to the **atomic size factor**, it has been found that primary solid solution is restricted to ca 1% when the energy of the solid solution exceeds $4k_BT$ per atom. If it is assumed that a misfit solute atoms behaves as and elastic sphere in a hole of the wrong size in the metal matrix, it can be shown[24] that the strain energy E is given approximately by

$$(4.28) \qquad E = 8\pi\mu r_0{}^3 \varepsilon^2 ,$$

where μ is the shear modulus of the metal, ε is the fractional radius variation of a solute atom relative to the radius of a solvent atoms, and r_0 and $(1 + \varepsilon)r_0$ are the strain-free radii of the solvent and solute atoms, respectively. Assuming the value of 0.15 for ε that we introduced in Section 4.6.1, and since μr_0^3 is approximately 0.7 eV for a number of common metals, E evaluates as 0.40 eV per atom. At 1000 K, $4k_B T$ is 0.34 eV per atom, so that solid solution would be restricted under these conditions.

Hume-Rothery showed that the greater the difference in **electronegativities** $|\Delta \chi|$ (see Section 3.3.1) for the species in a binary alloy, the greater the tendency for electrons to concentrate in the more electronegative species. For small values of $|\Delta \chi|$ ordered solid solutions tend to be formed, but where $|\Delta \chi|$ is large intermetallic compounds, such as Mg_2Si, are formed. Intermetallic compound formation is an alternative to random solid solution, and primary solid solution is small in these alloys.

A particular phase occurs at a critical **electron/atom ratio** in certain alloys, such as those between copper and zinc, which can be associated with particular points in the filling of the Brillouin zones. When a solvent metal, such as copper, silver or gold, is alloyed with a metal of higher nominal valency, such as zinc, aluminium or tin, the boundary of the primary face-centred cubic α-phase from the solid solution is reached at an electron/atom ratio (epa) of 1.4. For the body-centred cubic β-phase it is 1.5, and for the more complex γ and ε phases the epa ratios are higher still, at 1.6 and 1.75.

These empirical observations of Hume-Rothery were explained later by band theory. The Fermi surface expands with increasing epa and approaches the boundary of a Brillouin zone, leading to an increased density of states $g(E)$ at the boundary. This situation results in a large band-energy with consequent increase in the thermodynamic stability of the phase. By representing the Fermi surface as a sphere in the free-electron approximation, it has been calculated that the sphere touches the first Brillouin-zone boundary at an epa of 1.36 for face-centred cubic, 1.48 for body-centred cubic, 1.54 for the γ-phase and 1.69 for the ε-phase.

Consider a binary alloy with phases 1 (α) and 2 (β). Figure 4.16 shows the variation of the density of states function $N(E)$ with energy E. Let the solvent metal be monovalent, with phase 1 as its primary solution. Its electron concentration is too low for the Fermi level to reach A. As increasing amounts of a second (divalent) metal are added, the electron concentration and Fermi level increase. Initially the free-electron parabola is followed, and both phases are favoured equally so far as the Fermi energy of the electrons is concerned. Beyond A, the $N(E)$ curve for phase 1 rises to a peak at B, as the zone boundary for this phase is reached; that for phase 2 is still following the free-electron parabola. In the given range of electron concentration, the Fermi energy of electrons in phase 1 is lower than that in phase 2, so that phase is thermodynamically the more stable. Beyond B the position reverses rapidly, as $N(E)$ falls off for phase 1 while that for phase 2 rises. At compositions that take the Fermi level just beyond B, the Fermi energy rapidly becomes unfavourable for phase 1, so the phase boundary for phase 1 lies at B. Similarly, the best composition for phase 2 is that which puts the Fermi level just at C.

Let phase 1 be the α-phase of the Cu—Zn system, and phase 2 the corresponding β-phase, and let the electrons be free enough for the Fermi surface to be treated qs a sphere of radius k_F. From the de Broglie equation (see Section 4.3) and (4.14), we have

$$(4.29) \qquad E = \tfrac{1}{2}\, m_e v^2 = h^2/(2m_e\lambda) = h^2 k_F^2/(8\pi m_e) \,,$$

so that

$$(4.30) \qquad \lambda = 2\pi/k_F \,,$$

as indicated by (4.12). Thus, we obtain, from (4.13)–(4.15),

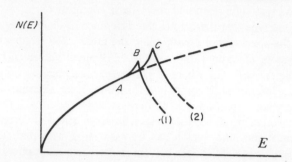

Fig. 4.16 Variation of the density of states function $N(E)$ with energy E for a binary alloy, such as Cu–Zn.

$$(4.31) \qquad \lambda = 2[\pi V/(3N)]^{1/3},$$

where V is the volume of a cubic unit cell, given by a^3, a being is the unit-cell dimension, and N has the meaning as before.

When the Fermi sphere just touches the Brillouin zone boundary at the nearest points, the corresponding wavelength is that for Bragg reflection, that is, $\lambda = 2d$, where d is an interplanar spacing. In a cubic F structure, the most widely spaced planes are $\{111\}$, for which $d = a/\sqrt{3}$. Also, $a = (4V/n)^{1/3}$, where n is the number of atoms in the volume V. Hence, we obtain

$$(4.32) \qquad \lambda = 2(4V/n)^{1/3} \sqrt{3}.$$

The epa ratio is N/n, which may be obtained by eliminating λ between (4.31) and (4.32):

$$(4.33) \qquad \text{epa} = \pi\sqrt{3}/4 = 1.36.$$

The most widely space planes in a cubic I structure are $\{110\}$. Remembering that there are two atoms per unit cell of side a, the corresponding epa evaluates to 1.48.

Table 4.6   Electron/atom ratios for some binary alloys

Phase	Molar composition	Valence electrons	Atoms	Electrons/atoms	
β	AgCd	1 + 2	2	3/2	(1.50)
β	Cu_3Ga	3 + 3	4	3/2	
β	Ag_5Sn	5 + 4	6	3/2	
γ	Ag_5Cd_8	5 + 16	13	21/13	(1.62)
γ	Cu_9Al_4	9 + 12	13	21/13	
γ	$Cu_{31}Sn_8$	31 + 32	39	21/13	
ε	$AgCd_3$	1 + 6	4	7/4	(1.75)
ε	$CuZn_3$	1 + 6	7	7/4	

We can see how these epa value apply to the phases in the silver—cadmium system discussed in Section 4.6.2. Additionally, Table 4.6, lists a number of alloys of species that are chemically dissimilar, but for which the formation of the β-, γ- and ϵ-phases are determined by the ratio of the number of valence electrons to the total number of atoms for the ideal composition.

When the filled energy states reach a Brillouin-zone boundary, it becomes energetically difficult to add further electrons. However, a change of crystal structure changes the Brillouin zone and its boundaries, so that an energetically favourable state can still be attained as more electrons are added. For a further study of the topic of alloys, the reader is recommended to the standard literature[24,25].

4.7 PROPERTIES ASSOCIATED WITH METALLIC SOLIDS

The metallic bond is non-directional, and metal structures have high coordination numbers. Metals are opaque and have a high reflecting power. Electrons near the Fermi energy level can absorb energy, so that their energy states are raised. If there is only a small interaction between these electrons and the lattice of positive ions, the energy absorbed is radiated away without change of phase, and the crystal is transparent to that radiation. Metals interact in this way with radiations of very short wavelength, such as X-rays.

Metals exhibit varying strengths, according to their structures. Deformation by gliding is common in metals, and it takes place most easily in directions parallel to close-packed planes of atoms. In the close-packed cubic structure type, there are four such planes, with Miller indices[5] (111), ($\bar{1}$11), (1$\bar{1}$1) and (11$\bar{1}$). In the close-packed hexagonal structure type there is only one such plane, (0001). Consequently, we find that the most malleable metals, such as copper, silver and γ-iron have the $A1$ structure type, whereas the harder and more brittle metals, such as beryllium, tungsten and α-iron, have the $A3$ or $A2$ structure types.

Metals have sharp but varied melting points (Hg 234 K, W 3683 K), and the liquid interval is long (Ga 2373 K, Hf 2375 K); their densities are also very variable (Na 971 kg m^{-3}, W 19.3 × 10^3 kg m^{-3}).

A study of groups 12–17 of the periodic table shows an increasing tendency towards metallic character with an increase in atomic number in each group. Thus, carbon (diamond) is covalent, tin is dimorphic with structures showing four-fold and six-fold coordination, and lead is metallic with the close-packed cubic structure type.

Conductors, semi-conductors and insulators

Metals are characteristic in their thermal and electrical conductivities. Figure 4.17 is a schematic arrangement of energy bands for metals, semi-conductors and insulators. If a band is partially filled, the species behaves as a metal and conducts electricity readily. If all bands are filled except for one or two which are either nearly full or nearly empty, the solid is a semi-conductor; and if all bands are filled the substance is an insulator. Semiconduction can arise also when all occupied bands are filled, but the energy gap between the uppermost filled, **valence** band and the next highest energy, **conduction** band is small. There are gradations between these classes that are revealed by measurements of electrical resistivity, as shown already by Table 3.2.

At 0 K, semi-conductors have a filled valence band separated from a conduction band by a small energy gap. *ca* 1 eV or less. Electrons cannot normally cross this gap, so that no electrical conduction will occur. When the temperature is increased, some electrons can cross over into the conduction band, whereupon electron flow can then take place if an electric field is applied to the material. The number of electrons promoted increases

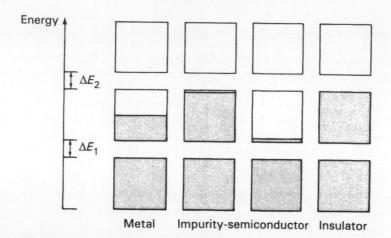

Fig. 4.17. Schematic illustration of energy bands in solids; the shading indicates electron occupancy. The energy gaps ΔE_1 and ΔE_2 are forbidden ranges for electron energies.

with increasing temperature, so that the electrical conductivity of a semi-conductor *increases* as the temperature is increased. Metals, however, behave in the opposite way: although an increase in temperature causes more electron excitation, there is also an increase in the vibrations of the lattice of positive ions, so that the electron-scattering, current-limiting

process increases and the steady current *decreases*.

In the conduction process of a semi-conductor, the conduction electrons lie in the tail of the Fermi-Dirac distribution, Figure 4.4, so that electrical conductivity follows a Boltzmann distribution. If E_v is the energy of the uppermost level of the valence band, then we can write

$$(4.34) \qquad\qquad \Delta E/2 = E_F - E_v \,,$$

whence, from (4.20),

$$(4.35) \qquad\qquad f(E) = \exp[-\Delta E/(2k_B T)] \,,$$

and the electrical conductivity σ ($= 1/\rho$) can then be expressed by an Arrhenius-type equation:

$$(4.36) \qquad\qquad \sigma = A \exp[-\Delta E/(2k_B T)] \,,$$

where A is a constant.

The number of charge carriers in a semi-conductor can be increased by **doping**, that is, by implanting impurity atoms at a concentration of *ca* 1 in 10^9 into highly pure material, and converting the pure, **intrinsic** semi-conductor into an **extrinsic** semi-conductor. If the impurity has fewer electrons than the host, such as gallium in silicon, they extract electrons from the conduction band leaving 'holes' that aid the movement of electrons. Such a material is a *p*-type semi-conductor, since the hole is *positive* with respect to electrons. If, instead, the dopant has more valence electrons than the host, such as arsenic in germanium, the additional electrons occupy otherwise empty bands and create *n*-type (*negative*) semi-conductors. Materials of ultra-high purity can be prepared by the technique of **zone refinment**[17].

In this technique, a cylindrically-shaped specimen is heated progressively along its length by an encircling heating element. As the liquid zone advances, the contaminants are partitioned into the liquid phase and accumulated. A succession of heating cycles leads to pure material at one end of the specimen and an impurity region at the other. Controlled quantities of a desired impurity may then be added to the pure material. The properties of doped materials often differ markedly from those of the pure material. Thus, iron containing 1-2% of carbon produces a brittle material (cast iron), whereas the pure material remains malleable and ductile down nearly to absolute zero.

ANSWERS TO EXAMPLES 4

4.1. (a) $C_v = 3R = 24.9$ J K^{-1} mol^{-1}. (b) $C_v = 1944(10/428)^3 = 0.0248$ J K^{-1} mol^{-1}.

4.2. $N/V = N/a^3 = k_F{}^3/3\pi^2$. More specifically, $N/V = Zn_e/a^3$, where Z is the number of atoms per unit cell, and n_e is the number of conduction electrons per atom. Thus, $N/V = (4 \times 1)/(0.361 \times 10^{-9})^3 = 8.5 \times 10^{28}$ m^{-3}. $E_F = [h^2/(8\pi^2 m_e)]k_F{}^2 = [h^2/(8\pi^2 m_e)](3\pi^2 N/V)^{2/3} = 1.13 \times 10^{-18}$ J $atom^{-1}$, or 7.05 eV $atom^{-1}$.

4.3. A typical tetrahedral hole is centred at ¼, ¼, ¼, and an octahedral hole at ½, 0, 0. Hence, the required distance is $(a^2/16 + a^2/16)^{1/2}$, or $a/\sqrt{8}$.

PROBLEMS 4

4.1. Calculate the packing fraction for a close-packed hexagonal array of identical spheres.

4.2. Calculate the Fermi energy and the Fermi temperature for lithium; the body-centred unit cell has a side a of 0.350 nm.

4.3. (a) Draw a two-dimensional square lattice in k-space, and mark in the boundaries of its first three Brillouin zones. (b) What is the first Brillouin zone for a lattice with a primitive unit cell of side a?

4.4. If a spherical Fermi surface just touches the first Brillouin zone in a primitive cubic unit cell, side a, of a metal, how many conduction electrons per atom are there in the metal?

4.5. How many tetrahedral holes and octahedral holes are there in *one* unit cell of the close-packed cubic structure type, and what are the fractional coordinates of the centres of the holes?

4.6. Cadmium sulfide CdS is a photoconducting material with an energy-band gap of 2.42 eV. Determine the largest wavelength of visible radiation that can excite a valence electron to the conduction band. What is the ratio of the resistivities of this material at 300 K and 400 K?

4.7. The resistance of a sample of germanium has been found to vary with temperature as follows:

T/K	300	350	400
R/Ω	20.4	2.67	0.581

Determine the band gap in both J and eV for this material.

4.8. The anharmonic vibration of atoms about their mean positions in a metal crystal may be represented by the potential energy function $V(r) = ar^2 - br^3$, where a and b are constants. By determining an expression for the mean displacement $\bar{r}$, show that $\bar{r}$ is directly proportional to temperature, that is, it is consistent with the solid expanding on heating. Note that, since r is small, $\exp(br^3) \approx 1 + br^3$.

4.9. How may one describe the crystal structure of gold in concise crystallographic terms?

4.10. The number of electrons of energy E_F in a metal that can change energy states as the temperature is raised above 0 K is[16] $g(E_F)k_BT$. These electrons interact with lattice vibrations and increase their energy by k_BT, so that the total increase in energy is $g(E_F)(k_BT)^2$. Determine an expession for the electronic contribution to the heat capacity at constant volume. How does the result interact with equation (4.17). If E_F for copper is 7.06 eV per atom, calculate the electronic contribution to the heat capacity at constant volume for copper at 25 °C.

4.11. Consider the coordination of six nickel atoms around arsenic in the nickel arsenide structure, Figure 4.15. If each nickel atom in this [AsNi$_6$] structural unit is joined to its *three* nearest-neighbour nickel atoms, a **coordination polyhedron** for this structural unit is formed. What is the point-group symmetry of this polyhedron?

4.12. It has been reported that elemental polonium crystallizes in a primitive cubic lattice, with space group *Pm3m*, the only material known to do so. The unit cell side *a*, from X-ray diffraction, is 0.335 nm, and the relative atomic mass is 210.0. Calculate the density of polonium. What would the density be if it packed in a face-centred cubic unit cell without change of atomic radius?

CHECKLIST 4

At the end of this chapter, you should be able to:

1. Define core, conduction (valence) electron and electron gas;
2. Define classical solid and write the potential energy and kinetic energy terms for a classical oscillator;
3. Define heat capacity and use the Debye T^3 law for heat capacity at low temperatures;
4. Outline the wave-mechanical free-electron theory for metals;
5. Describe energy band, Fermi energy and *k*-space;
6. Describe the density of states and the Fermi-Dirac distribution functions;
7. Calculate the mean electronic energy at low temperatures;
8. Outline band theory;
9. Define Brillouin zone and its boundary features, and Wigner-Seitz cell;
10. Relate energy bands to molecular orbitals;
11. Describe the close-packed metal structure types, and calculate packing fractions;
12. Calculate metallic radii from the unit-cell dimensions of crystalline metals;
13. Describe the interstices (holes) in simple metal structures and give the fractional coordinates of their centres;
14. Describe simple binary alloys and apply Vegard's rule;
15. Classify binary alloys into convenient types;
16. Give examples of typical binary alloys, and describe superstructures (superlattices);
17. Describe Hume-Rothery's electron/atom rule for binary systems, and outline their theoretical basis;
18. Describe the structural and physical characteristics associated with the metallic bond.

5

Looking at molecular structures

5.1 INTRODUCTION

So far, we have discussed three of the four classes into which we have chosen to divide solids, and we turn our attention now to molecular compounds; Figure 5.1 is a schematic illustration of the interactions involved in the four classes of solids under discussion in this book. Here, we shall be concerned with species that are attracted one to the other by forces of a dipolar character.

There are several kinds of dipolar attraction, but we shall consider three types: the interactions between two dipolar species, between a dipolar and a non-polar species, and between two induced-dipolar species, often termed **Keesom** energy, **Debye** energy and **London** (dispersion) energy, resepctively. Sometimes, these and other similar interactions are designated collectively as **van der Waals** energy (or forces), since they arise in the departures of real gases from the van der Waals equation of state.

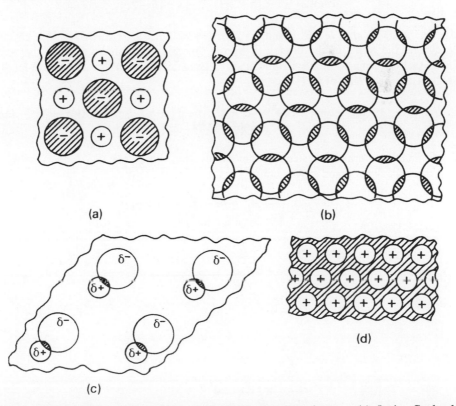

(a)

(b)

(c)

(d)

Fig. 5.1. Diagrammatic representation of interatomic bonding forces. (a) Ionic: Coulombic attraction between positive and negative ions. (b) Covalent: overlap of orbitals; 'sharing' of electrons. (c) van der Waals: overall dipolar (permanent or induced) attraction δ^+ and δ represent partial charges, which may be transient. (d) Metallic: positive ions in a 'sea' of electrons.

5.2 DIPOLE-DIPOLE INTERACTION

Two dipolar species attract each other, and an energy of interaction is set up between them. A complementary repulsion energy is set up because, under Earnshaw's theorem (in electrostatics), a system of charged particles under attraction alone cannot achieve a state of equilibrium. If dipolar molecules exist in a fluid or during the condensation process of forming a solid, then the molecules are able to rotate, and the field of one molecule acts to orientate the dipole of a neighbouring molecule. Generally, attractive forces dominate because they are of longer range than are forces of repulsion, and a nett attractive energy results.

Consider two molecules with permanent dipoles of moments p_1 and p_2 separated by a distance r, with a fixed orientation in one and the same plane (not the most general case), as shown in Figure 5.2. Let the effective charges at the ends of each dipole be Q_1 and Q_2, where Q is a numerical charge multiplied by the electronic charge e. We can simplify the calculation by assuming that the dipoles have the same length R by considering Q_2 modified as necessary with respect to Q_1. The potential energy V_{dd} is Coulombic, so that

$$(5.1) \qquad \begin{aligned} V_{d,d} &= [1/(4\pi\varepsilon_0)] \, (2Q_1Q_2/r - Q_1Q_2/AD - Q_1Q_2/BC) \\ &= [Q_1Q_2/(4\pi\varepsilon_0)] \, (2/r - 1/AD - 1/BC) \\ &= [Q_1Q_2/(4\pi\varepsilon_0 r)] \, (2 - \{1 + [R^2/r^2 + 2R\cos(\theta)/r]\}^{-\frac{1}{2}} \\ &\qquad\qquad - \{1 + [R^2/r^2 - 2R\cos(\theta)/r]\}^{-\frac{1}{2}} \\ &= (a/r)[2 - \{1 + [b + 2c]\}^{-\frac{1}{2}} - \{1 + [b - 2c]\}^{-\frac{1}{2}} \,, \end{aligned}$$

where $a = Q_1Q_2/(4\pi\varepsilon_0)$, $b = R^2/r^2$ and $c = R\cos(\theta)/r$. Since $r >> R$, that is, $[b \pm 2c] << 1$, we expand the terms in the braces $\{\}$ by the binomial theorem to include the second power, that is, terms no higher than R^2 and r^2. Generally $(1 \pm x)^{-\frac{1}{2}} = 1 \mp x/2 + 3x^2/8$, so that the right-hand side of (5.1) becomes

$$\begin{aligned} (a/r)[2 &- \{1 - b/2 - c + (3/8)(b^2 + 4c^2 + 4bc)\} - \{1 - b/2 + c + (3/8)(b^2 + 4c^2 - 4bc)\}] \\ &= (a/r)[b - 3b^2/8 - 3c^2]. \end{aligned}$$

We eliminate the terms in b^2 because they involve fourth powers of R (and r), so that by substituting back for b and c, we obtain

$$\begin{aligned} V_{d,d} &= Q_1Q_2/(4\pi\varepsilon_0 r)[\, R^2/r^2 - 3(R^2/r^2)\cos^2(\theta)] \\ &= -Q_1Q_2R^2/(4\pi\varepsilon_0 r^3)[3\cos^2(\theta) - 1]. \end{aligned}$$

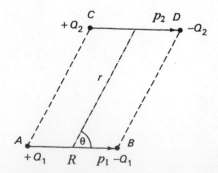

Fig. 5.2. Interaction of dipoles AB of moment p_1, charge Q_1 (q_1e) and length R, and CD of moment p_2, charge Q_2 and length R; r is the distance between their centres.

Remembering that a dipole moment p is $Q \times R$, we obtain finally

(5.2) $$V_{d,d} = -p_1p_2[3\cos^2(\theta) - 1]/(4\pi\varepsilon_0 r^3) .$$

The general case is that the two dipoles make angles θ_1 and θ_2 with the line joining their centres, and with ϕ defined as the angle of rotation of p_1 relative to p_2. Then, we would obtain

(5.3) $$V_{d,d} = -p_1p_2[2\cos(\theta_1)\cos(\theta_2) - \sin(\theta_1)\sin(\theta_2)\cos(\phi)]/(4\pi\varepsilon_0 r^3) .$$

In deriving (5.2), $\theta_1 = \theta_2 = \theta$ and $\phi = 0$. Maximum interaction occurs for either a collinear, head-to-tail orientation ($\theta_1 = \theta_2 = 0$), or an antiparallel orientation ($\theta_1 = 0$, $\theta_2 = \pi$); in these situations, we obtain

$$V_{dd} = \mp 2\, p_1p_2\,/(4\pi\varepsilon_0 r^3) .$$

In a fluid or in a solid that has condensed from a fluid-like arrangement of its component molecules, there will be a distribution of the angles θ_1, θ_2 and ϕ, so that we write an average form of the potential energy (5.3) as

$$\overline{V_{dd}} = -p_1p_2/(4\pi\varepsilon_0 r^3)\, \overline{f(\theta,\phi)}\ \overline{P(\theta,\phi)} .$$

If we assume a zero value for ϕ, as in Figure 5.2, but allow all values of the angle θ, then $\overline{f(\theta,\phi)}$ is the value $[3\cos^2(\theta) - 1]$ developed above. For $P(\theta,\phi)$ we use the Boltzmann distribution dependent on the energy V_θ, where V_θ is taken from (5.2), so that $P(\theta,\phi) = \exp(-V_\theta/k_BT)$. Since $V_\theta \ll k_BT$, we can expand the exponential function to two terms, thus taking $P(\theta,\phi) \approx [1 - V_\theta/(k_BT)]$.

Hence, and substituting for $\overline{V_\theta}$, the average energy becomes

$$\overline{V_{dd}} = [-p_1p_2/(4\pi\varepsilon_0 r^3)]\,\overline{[3\cos^2(\theta) - 1]}\,[1 - \overline{V_\theta}/(k_BT)]$$

$$= [-p_1p_2/(4\pi\varepsilon_0 r^3)]\overline{[3\cos^2(\theta) - 1]} - \{[-p_1p_2/(4\pi\varepsilon_0 r^3 k_BT)]\overline{[3\cos^2(\theta) - 1]^2}\} .$$

In order to obtain the average values of the functions of θ, we employ the general statistical result[16]

$$\overline{\theta} = \int \theta f(\theta)\, d\theta \,/ \int f(\theta)\, d\theta ,$$

where $f(\theta)$ here is the θ-function in the Maxwell-Boltzmann distribution of speeds, namely, $\sin(\theta)\, d\theta$, whereupon $\overline{\cos^2(\theta)}$ evaluates to $1/3$.

In a similar manner, $\overline{[3\cos^2(\theta) - 1]^2}$ is $4/5$, but a more detailed analysis[27,28] shows that the correct factor is $2/3$. Hence,

(5.4) $$\overline{V_{dd}} = -2p_1^2p_2^2/[3(4\pi\varepsilon_0)^2 k_B T r^6] .$$

The inverse r^6 factor means that the interactional energy of two dipoles will be small in magnitude; the inverse dependence upon temperature shows that the average dipole-dipole energy decreases as the temperature increases, because the orientation of dipoles is then more random.

EXAMPLE 5.1. A system of two dipoles, each of moment is 1 D, has a molecular separation of 0.37 nm. Calculate the average dipolar energy at 25 °C. The answer is at the end of the chapter.

The form of the dipole-dipole energy shown by (5.4) is the Keesom energy, to which we referred in Section 5.1.

5.3 DIPOLE- INDUCED DIPOLE INTERACTION

Let a polar species of dipole moment p_1 be placed at a distance r from a non-polar species of polarizability α_2. A general, albeit lengthy, calculation shows[27,28] that the field E_1, arising from the permanent dipole, at a distance r from it is given by

$$E_1 = p_1[3 \cos^2(\theta) + 1]^{\frac{1}{2}}/(4\pi\varepsilon_0 r^3) ,$$

where θ is angle between the direction of r and the dipole axis. A dipole is *induced* in the non-polar species; its moment is p_2 and the distance between its positive and negative induced charges, Q_2, is R. The energy w taken up by it is Coulombic in nature, and given by

$$w = \int_0^R E_1 Q_2\, dR = \int_0^R p_2 Q_2/\alpha_2\, dR = Q_2^2 / \alpha_2 \int_0^R R\, dR = Q_2^2 R^2/(2\alpha_2) = \alpha_2 E_1^2/2 ,$$

since $p_2 = \alpha_2 E_1$. The work w is the negative of the interactional energy, so that the dipole-induced dipole energy $V_{d,id}$ becomes

(5.5) $$V_{d,id} = -\alpha_2\, p_1^2[3 \cos^2(\theta) + 1]/[2(4\pi\varepsilon_0)^2 r^6] ,$$

or

(5.6) $$V_{d,id} = -\alpha_2'p_1^2[3 \cos^2(\theta) + 1]/[2(4\pi\varepsilon_0)r^6] ,$$

where α' represents the **volume polarizability** of dimensions m^3 (often quoted in dimensions of cm^3), related to α by

(5.7) $$\alpha' = \alpha/(4\pi\varepsilon_0) .$$

The maximum interaction in (5.5) arises for $\theta = 0$ (in-line or head-to-tail) and, because the average value of $\cos^2(\theta)$ is 1/3, the average energy is

(5.8) $$\overline{V_{d,id}} = -\alpha_2'p_1^2/(4\pi\varepsilon_0 r^6) .$$

The dipole-induced dipole interaction term in this form is the Debye energy.

EXAMPLE 5.2. The hydrogen chloride molecule has a dipole moment of 1.03 D. It interacts with a molecule of methane of volume polarizability 2.60×10^{-24} cm^3. What would be the average interactional energies at distances (a) 0.30 nm, and (b) 0.40 nm?. The answers are at the end of the chapter.

5.4 INDUCED DIPOLE- INDUCED DIPOLE INTERACTION

The London, or dispersion, energy is quantum mechanical in origin, but a simplified derivation is not inappropriate. Consider two noble gas atoms A and B: each has an electronic distribution about its nucleus, averaged over a time that is large compared to the period of fluctuation of the electron density, that is spherically symmetric, so that it is non-polar. However, at any instant in time an asymmetry exists in the distribution of electron density in an atom, say A, and it behaves as a dipole of moment p_A. The strongest effect will arise for a charge separation equal to the Bohr radius, so that

$$p_A = ea_0 .$$

The field arising from this dipole will polarize a neighbouring species B of polarizability α_B, leading to an interaction that is analogous to the dipole-induced dipole interaction. From (5.5), with $\theta = 0$, we have

$$V_{id,id} = -p_A{}^2\alpha_B/(4\pi\varepsilon_0)^2 r^6 = -(ea_0)^2\alpha_B/(4\pi\varepsilon_0)^2 r^6 .$$

As a given transient dipole changes its value and orientation, so the induced dipole will follow it. This correlation ensures that the multitude of transients do not average to zero, but give an overall attractive potential energy, which is a way of looking at the overall negative energetic result of induction.

Since the polarizability α' is a volume parameter, we write generally

$$\alpha \approx 4\pi\varepsilon_0 r^3 ,$$

which, for our analysis, we may re-cast as

$$\alpha_B \approx 4\pi\varepsilon_0 a_0{}^3 ,$$

and a_0 may be shown to be that distance at which the Coulombic energy $e^2/(4\pi\varepsilon_0 a_0)$ is equal to $2h\nu$, where ν is the frequency of zero-point motion, that is,

$$a_0 = e^2/(8\pi\varepsilon_0 h\nu),$$

whereupon, by eliminating e and a_0, we obtain

$$V_{id,id} = -2\alpha_B{}^2 h\nu/[(4\pi\varepsilon_0)^2 r^6] .$$

This expression has the correct form, but the precise analysis by London gives the result for identical species as

(5.9) $$V_{id,id} = -(3/4)\alpha^2 h\nu/[(4\pi\varepsilon_0)^2 r^6] ,$$

and between dissimilar species as

(5.10) $$V_{id,id} = -(3/2)\alpha_1\alpha_2 h\nu_1\nu_2/[(\nu_1 + \nu_2)(4\pi\varepsilon_0)^2 r^6] .$$

If we make the substitution $h\nu = I$, where I is the ionization energy, (5.10) becomes

(5.11) $$V_{id,id} = -(3/2)\ \alpha_1\alpha_2 I_1 I_2 / [(I_1 + I_2)(4\pi\varepsilon_0)^2 r^6]\ ,$$

and with volume polarizabilities ,

(5.12) $$V_{id,id} = -(3/2)\ \alpha'_1\alpha'_2 I_1 I_2 / [(I_1 + I_2) r^6]\ .$$

EXAMPLE 5.3. The volume polarizability of methane is 2.6×10^{-24} cm^3, and the ionization energy is 12.6 eV. Determine the dispersion energy for a distance of separation of 0.3 nm. The answer is at the end of the chapter.

The London energy exists for all materials in all states. In the noble gas species it is the sole energy of cohesion in condensed phases. In other solids, it is overshadowed by the other energies present, and Table 5.1 indicates the three van der Waals energies that we have discussed listed for several common substances.

5.5 INTERMOLECULAR POTENTIALS

At large distances of separation r, two molecules do not interact, so that their combined potential energy is zero. As they are brought closer together, an attraction develops, so that the potential energy decreases as r becomes smaller. At very small distances, however, electron repulsion causes the potential energy to rise steeply. The combination of these two situations results in a curve of the form of Figure 5.3, with a minimum at the equilibrium distance r_e. This potential energy function in the form

(5.13) $$V(r) = 4\varepsilon_{LJ}[(\delta/r)^{12} - (\delta/r)^6]$$

is known as the Lennard-Jones 12–6 potential. If we differentiate (5.13) and set the derivative to zero for $r = r_e$, where $V(r)$ is a minimum, we find that $r_e = 2^{1/6}\delta$, where δ is close to the collision diameter for any given species. Substituting for r_e in (5.13) shows that $V(r_e) = -\varepsilon_{LJ}$. We shall not explore this equation in detail, but just consider the case of argon. For this species, ε_{LJ} is *ca* $124k_B$ K and $\delta = 0.34$ nm, so $V(r_e) = 1.03$ kJ mol^{-1} for a pair of atoms. Just as we considered the multiplication of the energy of a pair of ions in an ionic crystal by the Madelung constant, so we must take account of the neighbours in the case of a crystal of argon, by means of **lattice sums**. The calculations are lengthy, and have been discussed in the literature[28]. For the face-centred cubic lattice, the sums are 12.132 for the r^{-12} term and 14.454 for the r^{-6} term, so that the Lennard-Jones energy equation for a face-centred cubic structure takes the general form

Table 5.1 van der Waals energies/kJ mol^{-1} for common species

	Keesom	Debye	London	Total	ΔH_{sub}
Ne	0	0	−0.6	−0.6	0.6
Ar	0	0	−2.03	−2.03	2.0
O_2	0	0	−1.7	−1.7	2.0
HCl	−0.79	−0.24	−4.02	−5.05	4.8
NH_3	−3.18	−0.37	−3.52	−7.07	7.1
H_2O	−8.69	−0.46	−2.15	−11.3	11.3

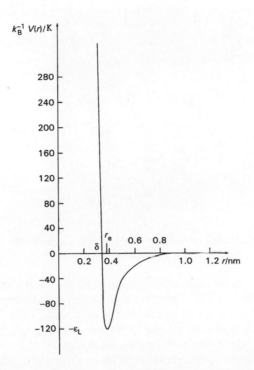

Fig. 5.3. Lennard-Jones 12-6 intermolecular potential energy function: $V(r) = 0$ at infinite r and at $r = \delta$; $V(r) = -\varepsilon_{LJ}$ at $r = r_e$ because $r_e = 2^{1/6}\delta$. For argon, $\delta = 0.34$ nm, so that $r_e = 0.38$ nm.

$$(5.14) \qquad V(r) = N\,4\varepsilon_{LJ}[12.132(\delta/r)^{12} - 14.454(\delta/r)^{6}],$$

where N refers to the number of atoms. The form $1/r^{12}$ for the repulsion is purely empirical, but the equation is nevertheless satisfactory for many simple solids.

5.6 MOLECULAR SOLIDS
The substances that may be termed molecular in the solid state can be classified for discussion according to the following scheme:

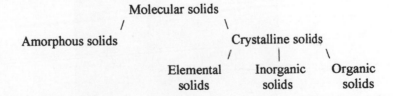

Molecular solids
Amorphous solids   Crystalline solids
Elemental   Inorganic   Organic
solids   solids   solids

5.6.1 Amorphous solids
Solids such as glass, plastics and resins are described as **amorphous**. Amorphous substances are metastable, although their change to crystalline forms may be a very slow process. For example, ancient glass may be found in a devitrified condition, in which some of the components have crystallized into more stable solid phases. The criterion of crystallinity is the appearance of a **spot pattern** under X-ray diffraction, indicative of a basic lattice structure for the solid.

Fig. 5.4. X-ray diffraction pattern from an amorphous solid. The characteristic concentric, diffuse maxima decrease in intensity as the angle of scatter increases (outwards).

Figure 5.4 is a typical X-ray diffraction photograph for an amorphous substance, such as silica glass (see Figure 3.12a). It shows a pattern of diffuse rings, indicating the nearest-neighbour distances, but no evidence of the long-range order associated with crystalline phases. Liquids show very similar diffraction patterns; a brief discussion of the structure of liquids has been given by the author elsewhere[16].

In Figure 5.5, we show the X-ray photograph of an ethene-propadiene polymer of approximately 70% crystalline character, drawn into a fibre. A spot pattern is superimposed on the diffuse rings, showing a repeating pattern in the (vertical) direction of the fibre axis.

The spot pattern of a fully crystalline material is shown in Figure 5.6. It is the cubic substance pyrite FeS_2, of space group $Pa3$ (point group $m3$). This sort of pattern (reciprocal lattice, q.v.) is typical of crystalline solids, and from similar data the detailed crystal structure can be elucidated[5].

Fig. 5.5. X-ray diffraction photograph of a drawn ethene-propadiene polymer at 70% crystallinity. The spot pattern is just discernible, and indicates a repeat period along the (vertical) fibre axis. The diffuse background ring pattern is characteristic of the amorphous state of matter.

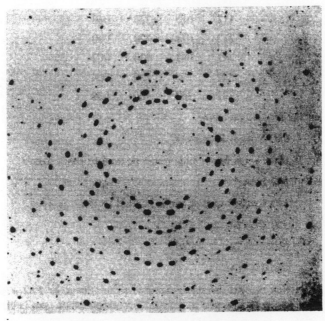

Fig. 5.6. X-ray diffraction photograph of FeS_2.

EXAMPLE 5.4. What symmetry is shown by Figure 5.6 and in what direction in the cubic crystal would it be encountered?

5.6.2 Packing of molecules

In the crystalline state, chemical moieties pack together so as to make the most energetically economical use of space. We have seen examples of this principle already in the structure of metals and ionic structures. Many organic compounds have molecules that are irregular in shape but which, nevertheless, achieve a maximum space-filling arrangement. Often this takes the form of a sort of 'lock and key' mechanism, in which the protrusion of one molecule fits into a recess in the adjacent molecule.

Figure 5.7 illustrates both good and uneconomical use of space with the same shaped molecule. The packing of organic molecules has been discussed in detail in the literature[29].

5.6.2 Elemental solids

We have considered certain elemental structures in previous chapters: the diamond form of carbon in Chapter 3, and certain metallic elements in Chapter 4. Here, we extend the discussion to include the structures of a number of other elements

Noble gases

The solidified gases neon, argon, krypton and xenon crystallize with the close-packed cubic structure $A1$, illustrated in Figure 4.8a, whereas helium adopts the close-packed hexagonal $A3$ structure type, Figure 4.8b. For them all, the coordination number is 12, with a packing efficiency of 74%. The noble gases are non-polar, so that the cohesive energy is the London energy expressed by (5.9), or the equivalent expression with the ionization energy term.

The small magnitudes of the London energy for the noble gases is evident from their very low melting points, 83 K in the case of argon. At this temperature, the mean kinetic energy $(3/2)k_BT$ is 1.035 kJ mol^{-1}, whereas the Lennard-Jones energy is –1.032 kJ mol^{-1}.

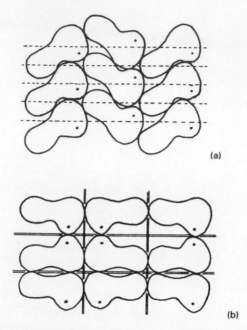

Fig. 5.7. Packing of irregular-shaped organic molecules. (a) Economical packing; 'lock and key' principle. (b) Uneconomical packing, leading to a higher percentage of void space compared with that in (a).

Thus, at 83 K the kinetic energy and the intrinsic energy are balanced, so that a solid can and does exist.

Other elemental structures

We discussed the diamond structure of carbon in Chapter 3. Another important allotrope of carbon is graphite, Figure 5.8. In this layer structure the C—C σ-bonds within each layer are covalent, formed from sp^2 hybrid orbitals, of length 0.142 nm, whereas the distance between the layers is 0.335 nm. The space group of graphite is $P6_3mc$ (see Figure 1.42), with the four atoms per unit cell in Wyckoff a and b positions. The London forces between planes are much weaker than those between the atoms in each plane. Hence, graphite shows a pronounced cleavage parallel to the basal (0001) planes, which is responsible for its properties as a lubricant. The thermal expansion of graphite is very anisotropic: the unit cell dimension c increases, while those of a and b are almost unchanged on heating.

The elements arsenic, antimony and bismuth, in periodic group 15, form puckered sheets of atoms, with each atom bonded to three others; Figure 5.9 illustrates the structure of arsenic. The σ-bonds in the layer are formed from hybrid s and p orbitals to give bond angles between 95° and 97°. The distances between the sheets increase from 0.333 nm in arsenic to 0.347 nm in bismuth. The coordination is distorted octahedral, with the three bonded neighbours being significantly closer (0.250 nm in As) than the three from an adjacent sheet (0.333 nm in As).

Elemental sulfur exists in several forms, an important allotrope being octasulfur S_8. Figure 5.10 is a stereoview of the space-filling 'crown' structure of this molecule. The S—S bond distance is 0.205 nm and the S–S–S bond angle 107.9°, with a nearest non-bonded distance of 0.369 nm. The molecule exhibits the non-crystallographic symmetry

$\bar{8}$ (point group $\bar{8}\,2m$), but it crystallizes in the orthorhombic space group *Fddd*. The unit cell contains sixteen S_8 molecules (128 atoms) in four sets of general positions (*h*).

By contrast, the structures of selenium and tellurium, in the same periodic group, form infinite helical chains, Figure 5.11. The covalent bond distances along the chains are 0.232 nm (Se) and 0.286 nm (Te), with bond angles of approximately 103°; the distances between the helical chains are 0.346 (Se) and 0.374 (Te), which are typical of

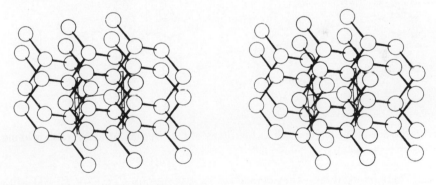

Fig. 5.8. Unit cell and environs of the graphite form of carbon. The space group is $P6_3mc$, with atoms occupying the Wyckoff *a* and *b* positions.

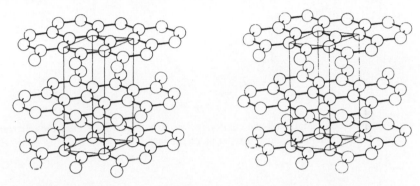

Fig. 5.9. Stereoview of the puckered-sheet structure of arsenic (also antimony and bismuth). The rhombohedral unit cell is shown by fine lines.

Fig. 5.10. Stereoview of a space-filling model of the S_8 molecule; point group $\bar{8}\,2m$.

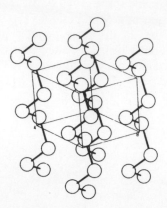 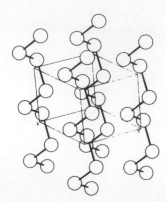

Fig. 5.11. Stereoview of the unit cell and environs of the structures of selenium and tellurium. The infinite chains are arranged in a helical manner along the directions of the z axis of their hexagonal unit cells.

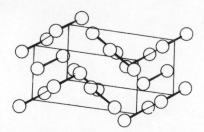 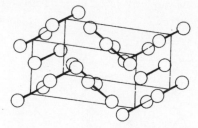

Fig. 5.12. Stereoview of the unit cell and environs of the structure of iodine (also chlorine and bromine): I–I = 0.268 nm; I_2–I_2 = 0.354 nm.

van der Waals interactions. They crystallize in space group $P3_121$ (or $P3_221$), with the atoms in Wyckoff *a* positions.

The halogens form diatomic molecules, and they are linked in the solid state by van der Waals forces. The atoms bond covalently at distance from 0.201 nm (Cl_2) to 0.268 nm (I_2). They crystallize in the orthorhombic space group *Cmca*, with the four molecules per unit cell lying in *m*-planes, at the Wyckoff *f* positions. Figure 5.12 shows the crystal structure of iodine. Fluorine exhibits a close-packed cubic structure, because the molecules are in free rotation and so achieve time-averaged spherical envelopes of motion.

The species that we have discussed, in the periodic groups 14 to 17 tend to follow an '18 – *n*' rule ('8 – *n*' with the earlier numbering of the periodic groups), but not without exceptions. In the structures of the elements, the tendency to pack closely can be discerned. Thus, in the structures of selenium and tellurium, for example, the 'elbow' of one helix fits into a recess in an adjacent chain.

In the sheet-type structures, an atom in one sheet lies above a 'hole' in an adjacent sheet. These modes of packing are adopted so as to limit the amount of voids, thereby leading to economical, minimum-energy space-filling configurations.

5.6.3 Inorganic molecular solids

We have touched on this class of compounds already, in discussing the transition from essentially ionic compounds, through layer-type structures to molecular compounds, as the degree of polarization increased among the species in the structure. There are many inorganic materials that form discrete molecules that are linked by van der Waals forces in the solid state. Examples are HCl, SO_3, N_2O, CO_2, SnI_4, SF_6, $HgCl_2$, and so on.

Carbon dioxide is cubic, with space group $Pa3$, as with the structure of pyrite FeS_2, and a unit cell dimension a of 0.5575 nm at 83 K. The carbon atoms occupy Wyckoff b and oxygen the c ($x = 0.110$) positions. We shall evaluate some molecular geometry for this structure by means of a problem at the end of the chapter.

Mercury(II) chloride is a linear molecular structure, similar to that shown by iodine. It crystallizes with the space group $Pnma$ (Figure 1.39), with the molecules in Wyckoff c positions on the m-planes. The molecules are arranged such that each mercury atom is coordinated by a total of six chlorine atoms, and each chlorine atom by three mercury atoms, thus reflecting the composition $HgCl_2$; the covalent Hg—Cl distance is 0.225 nm.

5.6.4 Organic molecular solids, I

In this section, we shall include all organic compounds whether or no they incorporate hydrogen-bonding, clathrate structures, charge-transfer compounds and structures that enhance their cohesive energy by π-electron overlap.

Among aromatic hydrocarbons, the effect of **π-electron-overlap** can be traced in the melting-points of these compounds. In Table 5.2, we list the melting-points for some aromatic hydrocarbons and for aliphatic hydrocarbons with the same number of carbon atoms. Because the molecular planes of the aromatic hydrocarbons are orientated in the solid state in such a way that the π-orbitals, normal to these planes, can overlap, extra delocalization energy and enhanced stability arise.

In the crystalline state of biphenyl ($(C_6H_5)_2$), for example, planar molecules are arranged in nearly parallel pairs, to facilitate π-electron overlap. The planarity of the molecule permits conjugation throughout the whole molecule, which results in the central C—C bond being 0.148 nm, significantly shorter than the standard single-bond length. In the gas phase, the minimum-energy conformation results in the planes of the two rings in biphenyl being at approximately 45° to each other. Figure 5.13 illustrates the crystal structure of biphenyl; the shortest intermolecular contact distances are 0.37 nm.

A typical **charge-transfer** compound exists in the structure of $C_6H_6 \cdot X_2$, where X is a halogen. In the case that X is chlorine, the compound $C_6H_6 \cdot Cl_2$ crystallizes in space group $C2/m$ with two molecules per unit cell. Two carbon atoms of each benzene molecule in the unit cell lie on two-fold axes, Wyckoff positions h; the remaining carbon

Table 5.2 Melting points and relative molar masses of some hydrocarbons

Compound	T_m/K	M_r
C_6H_6	279	78.1
C_6H_{12}	280	84.2
$C_{10}H_8$	353	128.2
$C_{10}H_{18}$	230	138.3
$C_{14}H_{10}$	490	178.2
$C_{14}H_{24}$	335	192.4

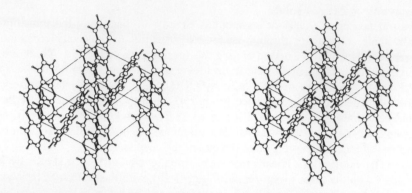

Fig. 5.13. Stereoview of the unit cell and environs of the structure of biphenyl $(C_6H_5)_2$; the space group is $P2_1/c$, with two molecules per unit cell.

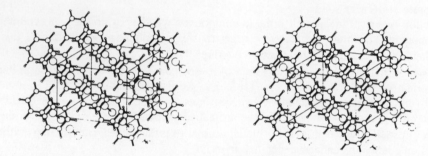

Fig. 5.14. Stereoview of the unit cell and environs of the structure of the benzene-chlorine charge-transfer compound $C_6H_6 \cdot Cl_2$; the circles in order of decreasing size represent Cl, C and H atoms. The shortest intermolecular contact distance is 0.328 nm, for Cl–ring centre.

atoms are in general positions. The chlorine atoms lie on m-planes, such that the Cl—Cl bond, of length 0.199 nm, lies normal to the benzene ring, with the shortest distance from Cl to the ring plane being 0.328. The structure is illustrated in Figure 5.14. In these structures, the halogens act as Lewis acids: the π-electrons of benzene are donors of electrons to the σ^* molecular orbitals on the halogens. These movements of electrons are charge-transfer transitions, and they are revealed in the UV absorption spectra of these compounds.

Compounds, in which one species is trapped by the structure of another and exist only in the solid state are known as **clathrates**. Often the energy of interaction between the host and the occluded species in the structure is small, with the host structure acting as a cage to the trapped molecule. A well known host structure is that of 1,4-dihydroxybenzene (quinol), and this compound acts as a mechanical trap for various small molecules, such as sulfur dioxide, methanol and cyanomethane. It is interesting to note that the dielectric constant for the quinol-cyanomethane is small, whereas that for quinol-methanol is large. We deduce that in quinol-cyanomethane, the occluded molecules are disordered, either statically or dynamically, whereas in quinol-methanol the methanol molecules are locked into position by hydrogen-bonding, between the quinol and methanol –OH groups .

The complex formed between nickel(II) cyanide and ammonia form a host structure for occluding molecules of benzene; the compound corresponds to the composition

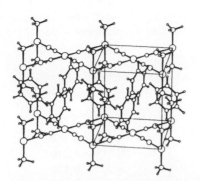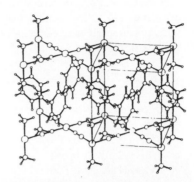

Fig. 5.15. Stereoviews of the unit cell and environs of the nickel(II) cyanide-ammonia-benzene clathrate structure $Ni(CN)_2 \cdot NH_3 \cdot C_6H_6$; the circles in order of decreasing size represent Ni, N, C and H atoms. The shortest contact distances are C (benzene)$\cdots$C (CN group) = 0.36 nm.

$Ni(CN)_2 \cdot NH_3 \cdot C_6H_6$, and the structure is shown by Figure 5.15. The nickel cations are coordinated octahedrally to four cyanide groups and two ammonia molecules, forming a cage with tetragonal symmetry that retains the molecules of benzene. The space group is $P4/m$, with two formula-entities per unit cell. The nickel atoms occupy Wyckoff a and c positions, the cyanide groups are on m-planes in j positions, with $x \approx y$, two carbon atoms of the benzene molecule are on two-fold axes, Wyckoff i, with the remainder in general positions. A number of similar structures is known in which benzene is replaced by thiophene, pyrrole, furan or pyridine. These molecule are all of comparable size and can replace benzene without disturbing the geometry of the host.

5.6.5 Organic molecular solids, II
We consider now organic solids in general. The number of compounds involved here is vast, and we shall attempt a form of classification so that the discussion shall be of a manageable length. We may anticipate that the classification will not be without anomalies, but it will, nevertheless, aid the study of this aspect of solid state chemistry.

Classification of organic compounds
We may divide organic compounds into classes according to their basic shape, **equant, flat** and **long** molecules. In each of these classes, we can distinguish between compounds that are **apolar** and those that are **polar**. Among the numerous compounds that have now been analysed, we identify a constancy in the size and shape of certain structural units that is brought about by the characteristic values of bond lengths and angles that arise in the following way. If we consider the structure of the water molecule again, we know that the bond lengths are 0.096 nm and the bond angle 104.4°. If one hydrogen atom is replaced by the methyl group to form methanol, the remaining O—H bond length is not altered significantly in value. Again, we know that the average non-bonded contact distance between carbon atoms, over a wide spectrum of compounds, is approximately 0.37 nm. It transpires that this value can be correlated with the size of a carbon $2p$ orbital envelope that encompasses 95% of the electron density of that orbital. Results of this nature give a theoretical basis to the sets of expectation values for bond lengths and bond angles, and for non-bonded contact distances, often called van der Waals distances, or radii. These data have considerable structural significance, and we list a number of currently accepted, normal values in Tables 5.3 and 5.4. In these tables, we use the notation C_n to indicate a carbon atom bonded to n other atoms, so that C_4—C_3

means a single bond between a carbon bonded to three other atoms (sp^3) and one bonded to two other atoms (sp^2). In a similar way, the apex atom in a bond angle is designated, so that C_2 means an angle between a carbon atom bonded to two species with the carbon atom at the apex of the angle.

Table 5.3 Expectation values for bond lengths/nm and angles/°

Bond lengths – single bonds

H—H	0.074		C_3—C_2	0.145
C_4—H	0.109		C_3—N_3	0.140
C_3—H	0.108		C_3—N_2	0.140
C_2—H	0.106		C_3—O_2	0.136
N_3—H	0.101		C_2—C_2	0.138
N_2—H	0.099		C_2—N_3	0.133
O_2—H	0.096		C_2—N_2	0.133
C_4—C_4	0.154		C_2—O_2	0.136
C_4—C_3	0.152		N_3—N_3	0.145
C_4—C_2	0.146		N_3—N_2	0.145
C_4—N_3	0.147		N_3—O_2	0.136
C_4—N_2	0.147		N_2—N_2	0.145
C_4—O_2	0.143		N_2—O_2	0.141
C_3—C_3	0.146		O_2—O_2	0.148

Bond lengths – double bonds

C_3—C_3	0.134		C_2—O_1	0.116
C_3—C_2	0.131		N_3—O_1	0.124
C_3—N_2	0.132		N_2—N_2	0.125
C_3—O_1	0.122		N_2—O_1	0.122
C_2—C_2	0.128		O_1—O_1	0.121
C_2—N_2	0.132			

Bond lengths – triple bonds

C_2—C_2	0.120		C_3—C_3	0.140
C_2—N_1	0.116		C_2—N_2	0.134
N_1—N_1	0.110		N_2—N_2	0.135

Bond angles

Apex atom	Geometry	Example	
C_4	Tetrahedral	CH_4	109.5
C_3	Planar	C_2H_4	120
C_2	Bent	OHC–	109.5
	Linear	HCN	180
N_4	Tetrahedral	NH_4^+	109.5
N_3	Pyramidal	NH_3	107.5
	Planar	H_2NCHO	120
N_2	Bent	H_2CHN	109.5
	Linear	HNC	180
O_3	Pyramidal	H_3O^+	190.5
	Bent	H_2O	104.4

Table 5.4 van der Waals radii/nm for some common species

H	0.12	C	0.185
CH_3	0.20	Si	0.21
N	0.15	P	0.19
As	0.20	Sb	0.22
O	0.14	S	0.185
Se	0.20	Te	0.22
F	0.135	Cl	0.18
Br	0.195	I	0.215
C_6H_5 thickness	0.37	H_2O	0.14

In the case of the van der Waals radii, we may find, on the one hand, that two non-bonded atoms are further apart than the sum of their van der Waals radii, because steric factors prevent the normal closet approach distance from being realized On the other hand, certain non-bonded distances may be observed to be significantly less than the expectation values; in such cases, another interaction, such as hydrogen-bonding, will be present. Large structural units are known among organic compounds. For example, the

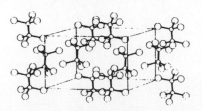 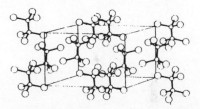

Fig. 5.16. Stereoview of the unit cell and environs of the crystal structure of hexachloroethane C_2Cl_6. (As before, in the following examples, the circles in order of decreasing size represent the species present in order of descending effective radii.) The space group is *Pnma* (see Figure 1.39), with four molecules of symmetry $\bar{3}\,m$ per unit cell situated on *m*-planes (Wyckoff *c*): C–C = 0.154 nm, C–Cl = 0.177 nm, mean bond angle = 110° ; closest intermolecular contact distance Cl...Cl = 0.37 nm.

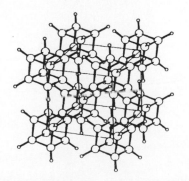 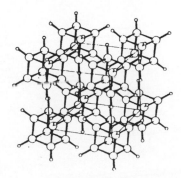

Fig. 5.17. Stereoview of the unit cell and environs of the crystal structure of cubane C_8H_8, space group $R\bar{3}$, with one molecule of C_8H_8 per unit cell; two C–H lie on the $\bar{3}$ axis at Wyckoff *c* positions: C–C = 0.1549 nm and 0.1553 nm; C–H = 1.01 nm and 1.11 nm; C–C–C = 89.3° and 90.5°, C–C–H = 123°–127°; closest intermolecular contact distance = 0.37 nm.

Equant-shaped molecules

Apolar, equant-shaped molecules, such as methane, ethane, hexachloroethane (Figure 5.16), adamantane and cubane (Figure 5.17), tend to form closely-packed structures. In some structures, free rotation of the molecule is present, so that high symmetry is observed. For example, methane crystallizes in space group *Fm*3*m*, with the carbon atoms in Wyckoff *a* positions. Polar, equant species form a more open packing, often with hydrogen-bonding prominent in the the structure. Examples are found with urea, methanol, pentaerythritol (Figure 5.18) and oxalic acid dihydrate (Figure 5.19).

Flat-shaped molecules

Apolar, flat-shaped molecules tend to pack in the solid state with their planes approximately parallel one to the other. The staggered configuration is often adopted where π-electron overlap can be facilitated, as with biphenyl (Figure 5.13); other examples are shown by benzene (Figure 5.20) and anthracene (Figure 5.21). The experimental results for benzene illustrate the characteristic difference between bond

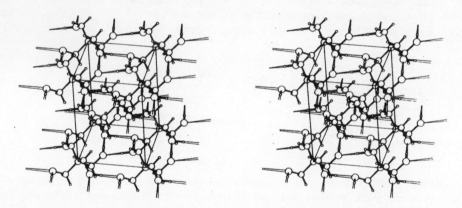

Fig. 5.18. Stereoview of the unit cell and environs of the crystal structure of pentaerythritol $C(CH_2OH)_4$. The space group is $I\bar{4}$, with two molecules per unit cell; the double lines are hydrogen bonds. We study this structure further by means of a problem.

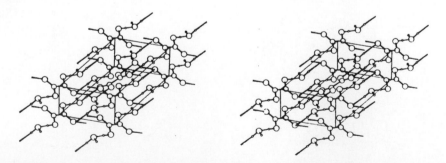

Fig. 5.19. Stereoview of the unit cell and environs of the crystal structure of oxalic acid dihydrate $(CO_2H)_2 \cdot 2H_2O$: space group $P2_1/n$ ($\equiv P2_1/c$), number of molecules per unit cell 2 is four, so that they le on centres of symmetry in the space group; C–C = 0.153 nm, C=O = 0.119 nm, C–O = 0.129 nm. The double lines represent hydrogen bonds, the shortest (strongest) of which is >C=O...HOH, 0.249 nm; the sum of the van der Waals radii is 0.325 nm.

lengths from X-ray diffraction, which measures distances between electron density maxima, and neutron diffraction, which pin-point the atomic nucleii. If the molecules possess polarity, the packing is modified by dipolar and hydrogen-bonded interactions, as in 1,4-dinitrobenzene (Figure 5.22) and 4-nitrophenol (Figure 5.23).

Long-shaped molecules

Long, apolar molecules tend to lie in parallel or staggered configurations, in order to achieve good space filling. This type of structure has been observed with long-chain,

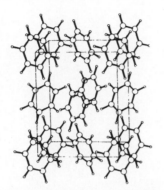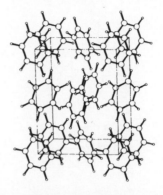

Fig. 5.20. Stereoview of the unit cell and environs of the crystal structure of benzene C_6H_6. The space group is *Pbca* with the species in general positions (Wyckoff *c*); C–C = 0.1398 nm, C–H 0.1077 nm (X-ray diffraction), but 0.1392 nm and 0.1090 nm (neutron diffraction).

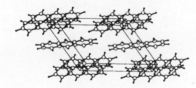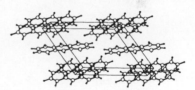

Fig. 5.21. Stereoview of the unit cell and environs of the crystal structure of anthracene $C_{14}H_8$: space group $P2_1/a$ ($\equiv P2_1/c$), $Z = 2$, with the molecules on centres of symmetry. The molecules are planar, with C–C ranging from 0.1366 nm to 0.1436 nm, the longer values applying to the ring junctions. The precision of the results is high, so that the differences are real.

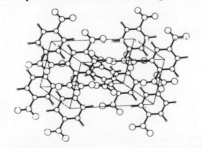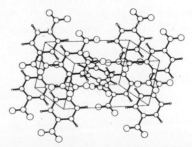

Fig. 5.22. Stereoview of the unit cell and environs of the crystal structure of 1,4-dinitrobenzene $C_6H_4(NO_2)_2$: space group $P2_1/n$; $Z = 2$, with the molecules in special positions on centres of symmetry. The molecules are non-planar, the angle between the planes of the ring and an –NO$_2$ group being *ca* 9°; C–C = 0.138 nm, C–N = 0.148 nm and N–O = 0.121 nm.

aliphatic hydrocarbons, such as octane (Figure 5.24), hexane and dicyanoethyne; the latter two structures are illustrated under Problem 5.7 (Figures P5.1 and P5.2, respectively). If a species possesses a dipole moment, there arises a tendency for molecular association in pairs through dipolar and hydrogen-bonded interactions. Long-chain alkylammonium salts are ionic, and frequently exhibit orientational disorder, which we shall discuss briefly in Section 5.8. Examples of long, polar molecules are decanamide (Figure 5.25), apidic acid $HO_2C(CH_2)_4CO_2H$, potassium caprate $CH_3(CH_2)_8CO_2^-K^+$ and and propan-1-ammonium chloride $CH_3(CH_2)_2NH_3^+Cl^-$. Though the latter two substances contain ions, they are best represented in this structural class.

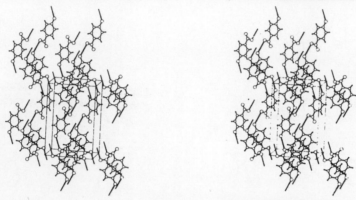

Fig. 5.23. Stereoview of the unit cell and environs of the crystal structure of 4-nitrophenol $C_6H_4OH(NO_2)$: space group $P2_1/n$ ($\equiv P2_1/c$), $Z = 4$, with planar molecules in general positions; bond lengths have standard values, and intermolecular $-O-H...O$ distances are ca 0.282 nm.

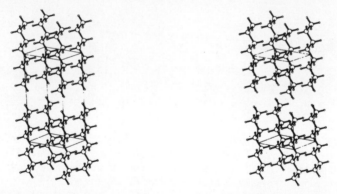

Fig. 5.24. Stereoview of the unit cell and environs of the crystal structure of octane C_8H_{18}: space group $P\bar{1}$. The C–C bond lengths range from 0.151 nm to 0.154 nm, and C–C–C bond angles are $112 \pm 0.2°$.

Fig. 5.25. Stereoview of the unit cell and environs of the crystal structure of decanamide $CH_3(CH_2)_8CONH_2$. The space group is $P2_1/a$, with atoms in general positions: C–C = 0.147 nm, C–N = 0.131 nm, C–O = 0.128 nm; intermolecular $-O-H...O$ distance = 0.289 nm.

5.7 PROPERTIES ASSOCIATED WITH MOLECULAR SOLIDS

The molecular, or van der Waals, bonds are spatially non-directional, and can link an atom to an indefinite number of neighbours. In the noble gases, the London energy is the sole means of cohesion in the solid state. In other molecular compounds, relatively short covalent bonds act between atoms with longer-range interactions between molecules. Van der Waals energies depend upon polarizability and intermolecular distance, and in the case of the London energy also upon ionization energy. The dependence of London energy on polarizability is shown by the melting points of the silicon tetrahalides (Table 5.5): the polarizability of the halogen increases more rapidly than does the intermolecular distance along the series, so that the cohesive energy is increased in like manner. In a series of alkanes of corresponding molar masses, the melting point range is only 140 K.

Molecular solids form soft and brittle crystals, generally with low melting points and high thermal expansivities. Their electrical and optical properties are similar in the solid, liquid and solution because the electron systems of the component molecules do not interact strongly. They are non-conductors of electricity, and their physical properties generally tend to be markedly anisotropic. Good sources of structural data are readily available in the literature and elsewhere, to which the reader may wish to refer[33-36].

5.8 DISORDER IN SOLIDS

Many crystalline solids exhibit disorder by a random placement of one or more of their structural moieties in the unit cell. The disorder may be orientational, a static system, or it may arise dynamically, through free rotation in the solid state. We shall consider some examples so as to illustrate both types of disorder.

Sodium cyanide and potassium cyanide both exhibit the sodium chloride structure type, space group $Fm3m$, at room temperature (Figure 5.26). Sodium cyanide at 279 K and potassium cyanide at 233 K transform to an orthorhombic structure in which the cyanide ions are in fixed positions, Figure 5.27.

The sodium chloride structure can be taken up if the cyanide ions are either in free rotation, so that their time-averaged envelope of motion is spherical, or are in static, random disorder. The most recent investigations of these structures indicate the latter situation, with the axes of the cyanide ions lying mainly in the (100) and (111) planes of the orthorhombic structure. The transformation between the two structures of these cyanides has been used, together with lattice energies, to determine the charge distribution on the cyanide ion[16,30].

Complex halides of the type MAg_4I_5, where M is K, Rb or NH_4 have been studied[31,32] because of their high electrical conductivity in the solid state; specific conductivities of ca 1 Ω m^{-1}, which is comparable with that for a decimolar aqueous solution of potassium chloride, have been reported. They belong to space group $P4_132$ (or $P4_332$), with four formula-entities per unit cell. The solutions of the crystal structures of these

Table 5.5 Melting points, polarizabilities and van der Waals radii sums for the silicon tetrahalides

	SiF_4	$SiCl_4$	$SiBr_4$	SiI_4
M Pt/K	183	203	278	394
$10^{24}\alpha'/cm^3$	1.0	3.4	4.8	7.3
$\Sigma\,r/nm$	0.35	0.39	0.41	0.43

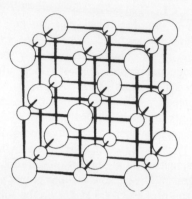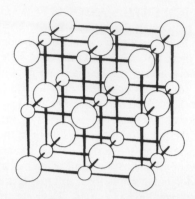

Fig. 5.26. Stereoview of the unit cell and environs of the cubic structure of sodium and potassium cyanides; the circles in order of decreasing size represent CN^- and Na^+/K^+. The space group is $Fm3m$, like that of sodium chloride, with four formula-entities per unit cell. Orientational disorder leads to an effectively spherical envelope for the CN^- ion when averaged over the many unit cells in a sample crystal.

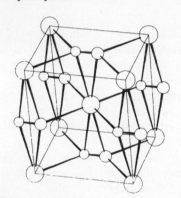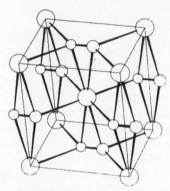

Fig. 5.27. . Stereoview of the unit cell and environs of the orthorhombic structure of sodium and potassium cyanides; the circles in order of decreasing size represent Na^+/K^+, N and C. The structure is orthorhombic with two formula-entities per unit cell; the cyanide ions are now in fixed positions. The relation to the cubic form is clear if we draw an orthorhombic unit cell of height c on a base of dimensions $|a + b|$ and $|a - b|$.

compounds are not wholly satisfactory, but it is evident that the sites for the silver ions have a fractional occupancy. It is probable that this form of disorder provides a mechanism for electrical conduction in the solid state by a process of positive ion migration.

Long-chain alkylammonium halides, such as $CH_3(CH_2)_3NH_3^+Cl^-$ exhibit dynamic disorder in the solid state. Their structures are tetragonal, with the chain along the four-fold symmetry axis c in the crystal. By X-ray diffraction, a repeat distance along c of approximately 0.13 nm was found. If we consider a planar zigzag chain of standard dimensions (see Table 5.3) in rotation about its axis, its envelope of motion becomes a cylinder of electron density with maxima at the carbon atoms. The distance between repeating maxima in this cylindrical envelope is 0.154 nm $\times$ sin(109.47/2) = 0.126 nm, which is in good agreement with the observed value. At low temperatures, these alkyl ammonium halides adopt monoclinic structures in which the lower symmetry is consistent with a static, planar carbon chain.

5.9 SOLUBILITY OF MOLECULAR COMPOUNDS

We have discussed solubility generally under ionic structures, in Section 2.3.6. Among ionic solids enthalpic effects are important, albeit not always dominant, because of the strong Coulombic forces involved in these compounds. The dissolution of molecular compounds is accompanied by less dramatic enthalpic changes, and entropy effects are more generally significant.

As with ionic solids, we are concerned with reaction between solute and solvent molecules. The free energy of dissolution, which is the algebraic difference between the free energy of solvation and the free energy of the crystal, determines the solubility of any molecular compound, but it is informative to consider the separate contributions of enthalpy and entropy.

We know that naphthalene dissolves in benzene but not in water, and that urea dissolves in water but not in benzene. Water is a hydrogen-bonded liquid: generally, hydrogen-bonded interactions are stronger than van der Waals interactions, especially when the polarity is small or absent in a solute species. Dissolution in water involves a disruption of some of the hydrogen-bonded structure of the solvent in order to accept and interact with the solute species.

In the pair naphthalene and water, there is no mechanism for interaction between solute and solvent that is sufficient to overcome the hydrogen-bonded interaction of the water molecules one for the other, so that the solute does not dissolve. In the pair naphthalene and benzene, however, the intermolecular attractions in the solvent are weak, so that enthalpic effects are very small and the increase in entropy of the dissolved solid, compared to that of the sum of the components, is a key factor.

Dissolution takes place because the system of molecules in solution represents a more random state than the system of solute and solvent separately. Thus, the entropy change is large and positive, so that from

$$\Delta G_s^{\circ} = \Delta H^{\circ} - T\Delta S^{\circ},$$

it follows that, for a negligible value of ΔH°, ΔG_s° will be negative, thus promoting dissolution of the naphthalene.

The situation is reversed with urea as a solute and benzene as solvent. There are no interactions between solute and solvent sufficient to break the van der Waals interactions between the molecules of urea in its solid state, especially as a dipole, of magnitude 4.6 D, exists in the molecule. In water, however, the polar groups of the urea molecules can interact with the water molecules, leading to a breakdown of the solid solute followed by its dissolution in the water. As with ionic solids, thermodynamic data on solubility lead to quantitative results that support both the arguments above and the experimental solubilities.

ANSWERS TO EXAMPLES 5

5.1. $\overline{V_{d,d}}/J$ = -2 × (3.3356)^4 ×10^{-120}/[16π^2 × 8.8542^2 × 10^{-24} × (0.37 × 10^{-9})^6 × 3 × 1.3807 ×10^{-23} × 298.15] = -6.31 × 10^{-22} J, or $\overline{V_{d,d}}$ = -0.38 kJ mol^{-1}.

5.2. $\overline{V_{d,id}}$ /J = -2.758 × 10^{-79} × L/r^6: (a) $\overline{V_{d,id}}$ = -0.23 kJ mol^{-1}; (b) $\overline{V_{d,id}}$ = -0.041 kJ mol^{-1}. Notice the rapid decrease in magnitude of the energy as r increases.

5.3. Using $I_1 = I_2$ and $\alpha_1 = \alpha_2$, $V_{id,id}$/J = -0.75 × 2.6^2 × 12.6 × 1.6022 × 10^{-25} / 0.3^6 = -1.40 × 10^{-20}, or $V_{id,id}$ = -8.5 kJ mol^{-1}. A useful comparison is the sublimation enthalpy of methane, 8.9 kJ mol^{-1}:

$$CH_4(s) \rightarrow CH_4(g) - \Delta H_s^{\circ}$$

PROBLEMS 5

5.1. The volume polarizability of argon is 1.66×10^{-24} cm, and its first ionization energy is 15.76 eV. What is the value of the London energy of two atoms at a distance apart of (a) 0.35 nm, and (b) 0.40 nm?

5.2. Compare the effect on the value of the Lennard-Jones 12–6 potential for xenon ($\delta =$ 0.37 nm, $\varepsilon_{LJ}/k_B = 229$ K) by changing the value of r_e by (a) +0.05 nm, and (b) –0.05 nm.

5.3. Using the data on solid carbon dioxide in the text. determine the C—O bond length and the nearest intermolecular distance. It may help to make a careful sketch plan of the structure.

5.4. Mercury(II) chloride has been listed with the non-standard space-group setting *Pbnm*, with $a = 0.4325$ nm, $b = 1.2735$ nm and $c = 0.5963$ nm. The standard setting is space group *Pnma*. Transform the unit cell dimensions to the new setting. It will help to draw the x, y and z axes, mark on the symmetry planes and the direction of translation of glide-planes, and then transform to the standard setting. Calculate the Hg—Cl distance in the original setting, given that the atoms occupy special positions as follow: $\pm(x, y, \frac{1}{4}; \frac{1}{2} - x, \frac{1}{2} + y, \frac{1}{4})$, with Hg (0.050, 0.126), Cl(1) (0.406, 0.255) and Cl(2) (0.806, 0.496).

5.5. Use the crystallographic data in the text relating to biphenyl to determine more features about the symmetry of the molecule and its crystal structure.

5.6. In addition to the data on pentaerythritol in the text, we have $a = 0.610$ nm, $c = 0.873$ nm, and the fractional atomic coordinates are as follow:

	x	y	z
C(1)	0	0	0
C(2), in –CH₂OH	0.162	0.123	0.097
O	0.314	0.247	0.002

Generate the remainder of the C and O coordinates for the molecule, and for a molecule distant a from it. Calculate C(1)—C(2), C(2)—O, and the shortest O…O intermolecular distance. You may have already written a program to calculate bond lengths. If not, the program *BOND* is available for such calculations (see Chapter 6). Compare the O…O distance with the sum of the van der Waals radii, and draw appropriate conclusions.

5.7. Compare the 'long' structures of hexane (Figure P5.1), intermolecular C…C contact 0.36 nm, and dicyanoethyne (Figure P5.2), intermolecular C…N contact 0.32 nm. What reason can be advanced to explain that dicyanoethyne does not pack in the simple manner shown by hexane?

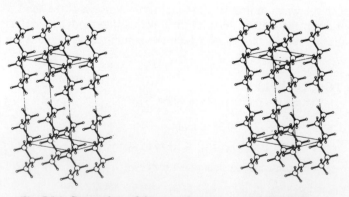

Fig. P5.1. Stereoview of the crystal structure of hexane C_6H_{12}.

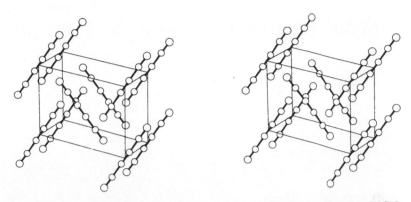

Fig. P5.2. Stereoview of the crystal structure of dicyanoethyne NC–C $\equiv$ C–CN

5.8. We have used the designations $P2_1/a$, $P2_1/c$ and $P2_1/n$ for one and the same space group. By drawing, or otherwise, show that $P2_1/a$ and $P2_1/n$ are equivalent to the standard setting $P2_1/c$.

5.9. Determine an expression for the energy of interaction between a dipole of moment p_1 and length R and a charge $+q_2$ distant r from the centre of the dipole. Hence, find the value of that energy when a charge of $+0.5e$ lies on the axis of a dipole of moment 1.5 D, 0.5 nm from its positive end.

5.10. Calculate the cohesive energies, in kJ mol^{-1}, of the noble gases neon and krypton in their solid states, given the following data.
Ne: Space group $Fm3m$; δ = 0.274 nm, ε_{LJ} = 35.7 K. Kr: Space group $Fm3m$; δ = 0.365 nm, ε_{LJ} = 163 K.

CHECKLIST 5
At the end of this chapter, you should be able to:
1. Understand the energy of interaction due to dipole–dipole (Keesom), dipole–induced dipole (Debye) and induced dipole–induced dipole (London);
2. Understand polarizability and volume polarizability;
3. Describe the Lennard-Jones 12-6 potential energy function;
4. Classify molecular solids;
5. Understand criteria for amorphous and crystalline natures of solids;
6. Describe some of the types of elemental solids, and packing considerations;
7. Describe inorganic molecular solids, and observe the relationship with this topic in Chapter 2;
8. Describe π-electron overlap, charge-transfer and clathrate compounds;
9. Classify organic molecular structures according to shape and polarity, and give examples of each class;
10. Appreciate 'standard' bond lengths and angles, and the reason for them;
11. Appreciate the near constancy of van der Waals radii of any given species over a range of molecules;
12. Describe the structural and physical characteristics associated with the molecular bond;
13. Understand some causes of disorder in solids;
14. Outline aspects of the solubility of organic compounds.

6

Problem solving: programs on the internet

6.1 INTRODUCTION

An important part of the study of chemistry is problem-solving. This activity may range from the evaluation of a parameter from a given equation to the detailed interpretation of a set of experimental data. An involvement with numerical work conveys an understanding of physico-chemical quantities. This facility is important, because computers and hand calculators produce sensible results only if they are supplied correctly with good data.

6.1.1 Solving numerical problems

Numerical problems give practice in relating experimental observations to models that have been set up from theoretical considerations. The insertion of magnitudes into a given equation is a common scientific activity that should be mastered.

The solving of problems leads to an appreciation of several important features:

(a) the orders of magnitude of physico-chemical quantities;
(b) the need for an understanding of units;
(c) the value of checking dimensional homogeneity;
(d) the sources of physico-chemical data;
(e) the precision of the data and its transmission to the result.

Most problems involve algebraic manipulation. However, it is essential to obtain a clear picture of the chemistry involved in a problem before embarking on a series of mathematical procedures, and it is useful to obtain an explicit algebraic expression before inserting numerical values. There are several advantages in so doing:

(f) the expression can be checked dimensionally;
(g) the possible cancellation of terms would simplify the arithmetic;
(h) similar problems with other magnitudes can be solved with only a little additional effort;
(i) it is good examination practice.

If the data are formed into an expression as the numbers 1 to 9 multiplied by the appropriate powers of 10, it is easy *and desirable* to estimate an approximate answer.

Suppose that we have for a relative permittivity ε_r

(6.1) $$(\varepsilon_r - 1) = Np^2/(9\varepsilon_0 k_B T) :$$

N (number of molecules per unit volume) $= 2.461 \times 10^{25} \text{ m}^{-3}$
p (dipole moment) $= 5.11 \times 10^{-30} \text{ C m}$
ε_0 (permittivity of a vacuum) $= 8.8542 \times 10^{-12} \text{ F m}^{-1} \text{ (C V}^{-1} \text{ m}^{-1})$
k_B (Boltzmann constant) $= 1.3807 \times 10^{-23} \text{ J K}^{-1} \text{ (C V K}^{-1})$
T (absolute temperature) $= 298.15 \text{ K}$

Inserting approximate quantities into (6.1) gives

(6.2) $\qquad (\varepsilon_r - 1) = \dfrac{2.5 \times 10^{25} \text{ m}^{-3} \times 25 \times 10^{-60} \text{ C}^2 \text{ m}^2}{10 \times 10 \times 10^{-12} \text{ F m}^{-1} \times 1 \times 10^{-23} \text{ J K}^{-1} \times 300 \text{ K}}.$

Equation (6.2) is dimensionally correct, because both sides are dimensionless. We can see that $(\varepsilon_r - 1) \approx 2 \times 10^{-3}$. Thus, when the expression is evaluated, we may write with confidence $(\varepsilon_r - 1) = 1.959 \times 10^{-3}$.

6.1.2 Suggested procedure

There are different ways of tackling problems, so that these notes are offered only as a guide. Sometimes a recommended stage may be changed or by-passed. Elegant derivations are often concise: the converse is not necessarily true, and failure to justify a stage in a derivation may indicate a lack either of judgement or of confidence. Over-elaboration of trivial detail and of arithmetic manipulation are equally unacceptable in a polished answer to a problem. Some degree of subjective judgement is involved in the process of problem solving, but it is expected that a correct numerical answer should include adequate evidence of the method used.

Thus,

(a) Read the problem carefully. If you think that it contains an ambiguity, assume the simplest interpretation of the ambiguity and comment on it.

(b) Summarize the given information by appropriate means, such as
    (i)   labelled drawings;
    (ii) energy-level diagrams;
    (iii) sketch-graphs, correctly labelled;
    (iv) defining symbols used in diagrams and formulae;
    (v) listing numerical values with units.

(c) Indicate relevant laws and equations that are employed in developing the problem, at least initially;

(d) State methods to be used; for example, 'take $\log_e$ of each side of equation (1)';

(e) When appropriate, formulate an explicit equation before inserting numerical data.

Look for cancellations of terms, and indicate any physical or functional approximations used;

(f) Do not make needless numerical approximations, but state any approximations that are made and include an estimate of the probable error, as far as is possible from the given data;

(g) For convenience, substitute a new symbol for a group of symbols in deriving an expression;

(h) Check the dimensions of equations for consistency. Remember that exponential, logarithmic and other arguments of functional are dimensionless;

(i) Insert numerical values into expressions carefully and determine an approximate result, as with (6.1) above;

(j) Think about the answer in terms of the chemistry of the problem. Check the results, especially if they seem doubtful;

(k) Keep a neat format in the answer to any problem.

6.1.3 Example problem and solution

The diffusion coefficient D of carbon in α-iron, as a function of temperature, is assumed to follow the equation:

(6.3) $$D = D_0 \exp[-E_d/(RT)] .$$

Values have been obtained for D as a function of temperature, T. It is required to verify the above equation, to find values both for the activation energy for diffusion E_d, with its estimated standard deviation, and for D_0, and to comment briefly on the results ($R = 8.3145$ J K^{-1} mol^{-1}). The following solution illustrate some of the above points.

The presence of R in the exponent, and its units, indicate that E_d is expected in J mol^{-1}; D_0 is the limiting value of D as $T \to \infty$. Taking natural logarithms (log$_e$, or ln) of each side of (6.3) gives

(6.4) $$\ln(D) = \ln(D_0) - E_d/(RT) .$$

If the graph of $\ln(D)$ against $1/T$ is linear (6.3) would be verified. The slope of the line, $\Delta \ln(D)/\Delta (1/T)$, is $(-E_d/R)$, and the intercept at $1/T = 0$ is $\ln(D_0)$. From the data we have:

$D/\text{m}^2 \text{ s}^{-1}$	$\ln(D/\text{m}^2 \text{ s}^{-1})$	T/K	$10^3/(T/\text{K})$
4.73×10	-46.800	300	3.3333
3.35×10	-33.330	500	2.0000
1.08×10	-27.554	700	1.4286
2.66×10	-24.350	900	1.1111
2.05×10	-22.308	1100	0.90909

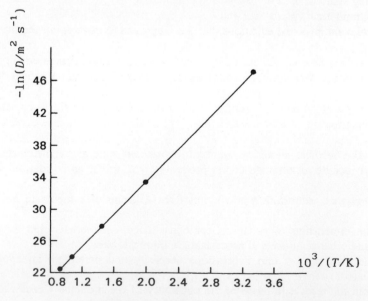

Fig. 6.1. Graph of $-\ln(D/\text{m}^2 \text{ s}^{-1})$ as a function of $10^3/(T/\text{K})$.

The straight-line graph (Figure 6.1) verifies (6.3) for the diffusion of carbon in α-iron over the experimental temperature range. By least squares, the slope of the line is -10103 K, giving E_d as 84.001 kJ mol^{-1}; $D_0 = 1.999 \times 10^{-6}$ m^2 s^{-1}. It is clear from the graph, and the least-squares correlation coefficient ($r = 1.0000$) that no datum merits exclusion from the calculation.

The sign and magnitude of E_d are reasonable, since work must be done on the system to bring about diffusion, and the energies of such processes are usually of the order of 1eV per atom (96 kJ mol^{-1}). The coefficient D increases with increasing temperature and, if (6.3) continued to hold, would tend to the limiting value 1.999×10^{-6} m^2 s^{-1}. However, at very high temperatures the material would melt, or even vaporize. The usefulness of D_0 here is in calculating D from (6.3).

The estimated standard deviation of the slope is (from least squares[16,39]) 8.5×10^{-3} K, which is transmitted to E_d as 0.007 kJ mol. The final value for E_d is (84.001 ± 0.007) kJ mol^{-1}, usually written as 84.001(7) kJ mol^{-1}.

6.2 COMPUTER METHODS IN PROBLEM-SOLVING

Throughout this book, reference has been made to least-squares minimization, calculation of Madelung constants and so on, procedures that often need to be invoked while gaining the desired familiarity with the subject matter. To this end a number of computer programs has been written that meet the requirements of problem-solving in the context of the book.

6.2.1 Computer programs

The set of programs available in conjunction with this text addresses procedures that are of importance to the subject of our study. Each of the programs that has been written is self-contained and mostly self-explanatory, and the programs may be executed on an *IBM*-compatible type of personal computer (*PC*), operating under *WINDOWS* or *DOS*. A monitor and a printer are needed in order to record the results from the programs; hard-disk working is assumed.

The programs are written in *FORTRAN*77, and are held as *.EXE* files that are compressed to a *.ZIP* file under the primary web reference **www.horwood.net/publish**, with *DOWNLOADS* as a secondary heading. The *.EXE* files may be decompressed by means of *WINZIP*7.0, which is available under *WINDOWS*. Click on the appropriate *DOWNLOAD* under the above web address, and then transfer the files to, say, *WINDOWS\DESKTOP* as, for example, *ADDONS.ZIP*. Click on this folder at the desktop, and decompress with *WINZIP*7.0, saving the *.EXE* files in an appropriate folder.

The programs are essentially *teaching aids*, and a study of the subject and of related topics is assumed. Each program is operated by entering its code name, for example, *<LSLI>*, for the linear least-squares program. Instructions appear on the monitor to guide the user through the program operation.

The convention has been adopted in these notes (and the programs) that data to be input are enclosed thus < >. Hence, the input of a title is shown as *<TITLE>*. Input items must be separated from one another either by a new line, or by one space or more if on one and the same line.

The programs display the instruction *KEY IN <X>* to indicate that the key X needs only to be depressed; the instruction *ENTER <X>* implies depressing the key X followed by depressing the *ENTER* key. If operation is carried out in a directory named *WORK*, a file named *INDATA* may be input as *<C:\WORK\INDATA>*, or simply *<INDATA>*. If the output file name is *<OUT>*, the results will go to the file *OUT* in the working directory, that is, in sub-directory *WORK*.

In most programs, *PAUSE* commands have been inserted so as to allow time for the information on the monitor to be read; execution is resumed by depressing the <Y> key, as directed on the monitor.

The data input may be from the keyboard or from a prepared file, or both, according to the nature of the program Where the input is lengthy, it is desirable to prepare and check a data file for this purpose. Some programs have been provided with an 'echo of data', to facilitate checking (and correction) of the input data.

Most of the programs contain data error traps, but not every possible data-error path has been explored, so that care should be exercised with the preparation and input of data. The output format of each program has been arranged to deal with a variety of problems. However, it must be remembered that the number of figures given in a result may not all be significant in relation to the data, and a correct appraisal of each result is essential.

The following notes and test data are given to ensure that the programs function correctly. If an apparent fault occurs, the user should check the input data carefully to ensure conformity with the program specification.

6.2.2 Linear least-squares program, *LSLI*
The least-squares program solves the equation

(6.5) $$y = ax + b$$

for the best values of a and b, assuming that the errors in x are negligible compared to those in y. The program also computes the root-mean-square error, the estimated standard deviations in a and b. and the Pearson correlation coefficient r; unit weights are assumed for the observations x.

After entering the code name <*LSLI*>, a number <0> or <1> is keyed in to signify that the data will be presented from the keyboard or a file, respectively.

A data set must contain at least three observations, and up to a total of 100 data is allowed by the program; the data format for n observations is as follows:

<*Title*>
<*n*.
<x_1> <y_1>
<x_2> <y_2>
<x_3> <y_3>
...   ...
<x_n> <y_n>

If the results, slope and/or intercept, are to be used to obtain physical quantities, as is quite common, then the estimated standard deviations σ in the slope and intercept may be transmitted to the derived parameters by procedures that are discussed elsewhere[16,39].

Test data

x	y
0.50	−0.511
1.00	−0.249
1.50	−0.002
2.00	0.252
2.50	0.511
3.00	0.747

Results
$a = 0.504$; $\sigma(a) = 0.0032$; $b = -0.758$; $\sigma(b) = 0.0063$; rms error $= 0.0067$; $|r| = 0.9999$.

6.2.3 Madelung constant program, *MADC*
This program for calculating a Madelung constant A addresses the

cubic (1)	hexagonal (2)
tetragonal (1)	trigonal (on hexagonal axes) (2)
orthorhombic (1)	monoclinic (2)

crystal systems; the numbers in parentheses are *system numbers* for this program. For hexagonal and trigonal crystals, a transformation must be applied such that the unique, 120° angle is called β, which means that the unit-cell constants are then entered in the order b, c, a, β, and the atomic coordinates in the order y, z, x.

The code name *<MADC>* is entered, followed by data entered at the keyboard <0> or a file <1>, in the following form:

Line 1: *<Title>*
Line 2: *<a> <b> <c> <β> <number of formula-entities per unit cell> <nearest-neighbour distance> <system number> <number n of atoms per unit cell> <summation limit>*
Lines 3 to $n+2$: $<x_1> <y_1> <z_1> <q_1>$

$$<x_1> <y_2> <z_2> <q_2>$$

$$.. \quad .. \quad ..$$

$$<x_n> <y_n> <z_n> <q_n>$$

a, b and c are in nm, and β is in deg; x, y and z are the fractional coordinates of the atoms ($x = X/a$, where X and a are in nm, and similarly for y and z) in *one complete unit cell*; $<q>$ are the charges on the ions, with their respective signs. The following data set is for the cesium chloride structure type.

Test data
<CsCl>
<0.4123> <0.4123> <0.4123> <90.0> <1> <0.35706> <1> <2> <3>
<0.0> <0.0> <0.0> < 1.0>
<0.5> <0.5> <0.5> <-1.0>

Results
$A(CsCl) = 1.762620$. The summation limit can take the integer values 2, 3, 4 or 5. The higher the limit the better the precision, but the longer the computation time:

Summation limit	A
2	1.762201
3	1.762620
4	1.762667
5	1.762663

In some structures, such as Rutile (TiO_2), where there is no unique nearest-neighbour distance, the shortest interatomic distance in the unit cell is used instead.

6.2.4 Electron-in-a-box, *BOXS*

This program calculates the wavefunctions and energies for the first eight solutions ($n = 1–8$) of the one-dimensional electron-in-a-box (free electron) wavefunctions. It can be applied also to conjugated molecular systems to predict spectral characteristics of excited states, by treating the π-electron system as a one-dimensional box.

On entering the code name *<BOXS>*, the program directs the user on the straightforward input to the program. In a conjugated system, it must be remembered that the potential energy function experienced by an electron does not actually increase to infinity at the end of the conjugated system (box), and precise results from this approximation should not be expected.

The results are given for $\psi(n)$ between 0 and the box length a (in nm) in integer steps from 0 to 24, that is, $\sin(n\pi x/a)$ is treated as $\sin[n(\pi/24)m]$, where m runs from 0 to 24, and n from 1 to 8.

The values of $\psi(n)$ are normalized to a maximum of 99, for convenience of output. The results show the form of the function sufficiently well to indicate the number of nodes for each value of n, and the field figures on the printed output can be sketched satisfactorily to show the forms of the wavefunctions.

Test data 1

For a box of length 1 nm, the energy difference for $n = 4$ to $n = 3$ calculates as 4.217×10^{-19} J.

Test data 2

For a conjugated system with three double bonds, with a mean bond length of 0.14 nm, the box length is 0.70 nm, and the first excitation energy calculates as 2.741 eV.

6.2.5 Point-group recognition program, *SYMH*

This program has been devised to assist in a study of the point-group symmetry of crystals and molecules, and a reference text [2,3] is suggested for concurrent study. The program uses a point-group recognition scheme based on a division of the point groups into four types, according to the presence of the symmetry elements m and/or $\bar{1}$ or neither, as discussed in the text.

The first step in the scheme is to search the given model for the presence of a centre of symmetry, a mirror plane, both or neither of these symmetry elements. Each given model should be then examined carefully, so as to identify the other symmetry elements present, the number of each of them, and their relative orientations.

The identification proceeds systematically in accordance with the programmed scheme: n refers to the maximum degree of pure rotational symmetry in the model, and N is the number of such axes. Where diads (two-fold rotation axes) are present with $n \geq 3$, there will be n of them in all.

Execution begins on input of the code name *<SYMH>*, followed by the model number and then questions about the symmetry of the model. If an incorrect response is given, it is noted by the program; subsequently, control is returned for further action to the point where the incorrect response was made.

Up to two incorrect attempts at identifying the point group are allowed; thereafter, a program break occurs and further study of the model, or of a new model, is suggested.

Crystal and molecular models used in conjunction with this program must be numbered in accordance with the schedule in Appendix 1 of this book.

Two exercises with the program form Example 1.2 in Chapter 1, and should be attempted, with others, during the study of that chapter.

6.2.6 Curve-fitting and interpolation, *INTP*

This program fits a curve to a series of data points x_1 y_1, x_2 y_2, ... x_n y_n. The program groups the data into ranges, and unique quadratic functions are fitted to each range, such that successive functions match in gradient and curvature.

Execution begins on entering the code name *<INTP>* followed by the data. Coefficients for the quadratic functions are derived and stored. The fitted curves must, of necessity, meet each data point, and other values are obtained by interpolation.

Values lying outside the extrema by less than 10^{-6} are considered as the respective extrema; numbers outside this range are rejected, because extrapolation may give unreliable results. The following data form a typical test set.

Test data
<Test>
<6>
<0> <7>
<1> <5>
<2> <9>
<3> <25>
<4> <59>
<5> <117>

Results
For $x = 4.4$, the interpolated value is 79.32.

6.2.7 Bond lengths, *BOND*

This program is very straightforward. The calling name is *<BOND>*, and the input consists of a *<Title>*, *<a>*, *b>*, *c>*, *<alpha>*,*<beta>*,*<gamma>*, *<number n of atoms>*, *<x>*, *<y>*, *<z>* up to a total of *n* atoms (*n* must not exceed 50). The first three lines of data are entered at the keyboard, and the *x, y, z* data are expected from a file named *XLDAT*. The results go to an output file named *DISTS*.

Test data
<Title Test>
<0.7> <0.8> <0.9> <90> <90> <90>
<3>
0.1011   0.2127   0.0079
0.2113   0.3014   0.0981
0.1515   0.2573 -0.1011

Results, on the output file *DISTS*:

Atom Numbers		Distance/nm
1	2	0.1326
1	3	0.1102
2	3	0.1875

Bond angles can be obtained by solving the appropriate triangle from the output results. The program can be used also to calculate lengths in reciprocal space, such as the magnitude of a direction [*UVW*], by using the reciprocal lattice constants a^*, b^*, c^*, α^*, β^*, γ^*.

6.3 DISCLAIMER

Although the greatest care has been taken in setting up and testing the programs described herein, the reader planning to use them should note that, from the legal point of view, there is no warranty, expressed or implied, that the programs are free from error or will prove suitable for a particular application. By using the programs the reader accepts full responsibility for the results produced, and the author and publisher disclaim all liability from any consequences arising from the use of the programs. The programs should not be relied upon for solving a problem whose incorrect solution could result in injury to a person or loss of property. If you do use the programs in such a manner, it is at your own risk. The author and publisher disclaim all liability from direct or consequential damages resulting from your use of the programs.

6.4 COPYRIGHT

The programs described herein are copyright by the author, Dr Mark Ladd, 1 April 1999. Readers of this book may use these programs freely, but they may not be sold, rented, or reproduced without the written permission of the copyright holder.

PROBLEMS 6

(*These problems are offered so that the reader may gain familiarity with the programs described above.*)

6.1. The molar polarization P_m of a substance is given by
$$P_m = (L/3\varepsilon_0)[\alpha + p^2/(3k_B T)].$$

The following data were obtained for ammonia:

T/K	292	309	333	387	413	446
$10^3 P_m/m^3\ mol^{-1}$	57.6	55.0	51.2	45.0	42.5	39.6

Use least squares to find the dipole moment p; then, given N—H = 0.107 nm and H–N–H = 107°, fine the charge on nitrogen and hydrogen in the ammonia molecule.

6.2.Calculate the Madelung constant for the fluorite CaF_2 structure, given the following structural data:
$a = 0.5464$ nm ; $Z = 4$; Ca^{2+} at $(0, 0, 0) + F$; F at $(¼, ¼, ¼; ¾, ¾, ¾) + F$.

6.3. Consider the symmetry of (a) the *inner tray* of an ordinary matchbox, and (b) a model of the trichloromethane molecule. Investigate their symmetry elements. Then, use the program *SYMH* to determine their point groups, given the model numbes 71 and 42, respectively.

6.4. The refractive index of pure water to light of wavelength 589.3 nm has the following values at the temperatures given. Find the value of n at 25 °C.

$T/$ °C	14	18	22	26	30
n	1.33348	1.33317	1.33281	1.33241	1.33192

6.5. Crystals of hydroxylamine NH_2OH are orthorhombic, with space group $P2_12_12_1$. The unit cell dimensions are $a = 0.7292$ nm, $b = 0.4392$ nm and $c = 0.4875$ nm; $Z = 4$. The nitrogen and oxygen atoms lie in general equivalent positions as follows:

$$x, y, z;\ ½ - x,\ \overline{y},\ ½ + z;\ ½ + x, ½ - y,\ \overline{z}\ ;\ \overline{x}, ½ + y, ½ - z.$$

For N: $x = 0.121$, $y = 0.244$, $z = 0.063$. For O: $x = 0.060$, $y = -0.062$, $z = -0.023$. Determine the unique intramolecular and intermolecular N—O distances. Indicate any that might be considered as hydrogen bonds.

CHECKLIST 6

At the end of this chapter, you should be able to:

1. Develop a good technique for solving problems;
2. Understand the value of obtaining an approximate numerical answer and of checking dimensions;
3. Use the least-squares procedure to fit a straight line to data;
4. Transfer standard errors in the least-squares constants to the related physical parameters;
5. Use computer programs to obtain practice in the additional procedures of calculating Madelung constants, calculating wavefunctions and energies for the electron-in-a-box model, recognizing point groups of molecular and crystal models, fitting a curve to data and interpolating from it, and calculating bond lengths from crystal structure data.

Appendix 1 Model numbers for the symmetry-recognition program *SYMH*, symmetry notation and example compounds

Point group	Model number/s	Example or possible example
1	91	CHBrClF, bromochlorofluoromethane
2	77	H_2O_2, hydrogen peroxide
3	84, 93	H_3PO_4, phosphoric acid
4	85, 94	$(CH_3)_4C_4$, tetramethylcyclobutadiene
6	88, 97	$C_6(CH_3)_6^{\cdot}$, hexamethylcyclohexadienyl
$\bar{1}$	78, 79	$C_6H_5CH_2CH_2C_6H_5$, dibenzyl
$\bar{3}$	48	$Ni(NO_2)_6]^{4-}$, hexanitronickelate(II) ion
$\bar{4}$	86, 95	$[H_2PO_4]^-$, dihydrogen phosphate ion
$\bar{6}$	89, 98	$C_3H_3N_3(N_3)_3$, 1,3,5-triazidotriazine
2/m	68–70,72–75, 80	CHCl=CHCl, *trans*-1,2-dichloroethene
4/m	56	$[Ni(CN)_4]^{2-}$, tetracyanonickelate(II) ion
6/m	37	$C_6(CH_3)_6$, hexamethylbenzene
$m\,(\bar{2})$	83, 92, 99	$C_6H_3Cl_3$, 1,2,4-trichlorobenzene
mm2	16, 64, 71, 76	C_6H_5Cl, chlorobenzene
3m	42	$CHCl_3$, trichloromethane
4mm	87, 96, 100	$[SbF_5]^{2-}$, pentafluoroantimonate(III) ion
6mm	81	$C_6(CH_2Cl)_6$, hexa(chloromethyl)benzene
222	67	C_8H_{12}, cycloocta-1,5-diene
32	43, 47	$[S_2O_6]^{2-}$, dithionate ion
422	55	$Co(H_2O)_4Cl_2$, tetraaquodichlorocobalt
622	36	$C_6(NH_2)_6$, hexaminobenzene
mmm	59–63, 65, 66	$C_6H_4Cl_2$, 1,4-dichlorobenzene
$\bar{6}\,m2$	44-46, 90	$[CO_3]^{2-}$, carbonate ion
4/m mm	49-54	$[AuBr_4]^-$, tetrabromoaurate(III) ion
6/m mm	29-35	C_6H_6, benzene
$\bar{4}\,2m$	57, 58	$ThBr_4$, thorium tetrabromide
$\bar{3}\,m$	38–41	C_6H_{12}, *chair*-cyclohexane
23	27	$C(CH_3)_4$, 2,2-dimethylpropane
m3	22–25	$[Co(NO_2)_6]^{3-}$, hexanitrocobaltate(III) ion
$\bar{4}\,3m$	17–21, 28	CH_4, methane
432	26	$C_8(CH_3)_8$, octamethylcubane
m3m	1–15	SF_6, sulfur hexafluoride

Noncrystallographic point groups

$\bar{8}\,2m$	-	S_8 sulfur
5	-	$C_5(CH_3)_5^{\cdot}$, pentamethylcyclopentadienyl
5m	-	C_5H_5NiNO, nitrosylcyclopentadienylnickel
$\overline{10}\,m2$	-	$(C_5H_5)_2Ru$, biscyclopentadienylruthenium
$\bar{5}\,m$	-	$(C_5H_5)_2Fe$, biscyclopentadienyliron
∞m	101	HCl, hydrogen chloride
∞/m	102	CO_2, carbon dioxide

Appendix 2 Calculation of Madelung constants

We give a simple method of calculating the Madelung constant for any structure, to a precision that is sufficiently high for most purposes. The method given here[11] has been applied successfully to a wide range of compounds. It assumes spherical charge distributions on the ions, with densities that are linear functions of the radial coordinate in reciprocal space. The Madelung constant A is given[12] by

$$\text{(A2.1)} \qquad A = d(g - Q)/rZ \, \Sigma_j \, q_j^2 - \pi r^2 d/V \, \Sigma_h \, |F_h|^2 \, \phi(h),$$

where the terms have the following meanings:

g	26/35;
Q	a correction for termination of the h series (as below);
d	a standard distance in the structure, often the nearest-neighbour distance;
r	an arbitrary distance, normally $0.495d$;
Z	the number of formula-entities in the unit cell;
q	the charge, including sign, of the jth species in the unit cell;
V	the unit-cell volume;
h	the magnitude of the reciprocal lattice vector h;
$\phi(h)$	$288[\alpha \sin(\alpha) + 2 \cos(\alpha) - 2]^2 \, \alpha^{-10}$
α	$2\pi hr$
F_h	$\Sigma_j \, q_j \exp(i2\pi h \cdot x_j)$; the crystallographic structure factor for point atoms of form factors q_j .

The sums over j include all atoms in the unit cell and the sum over h includes all reciprocal lattice vectors hkl in a sphere of radius α; h, k and l are the components of h with respect to the reciprocal lattice axes. The series termination correction Q depends on the radius α according to the following table:

α	Q	α	Q
2π	0.00030	4π	0.000012
3π	0.000090	5π	0.0000057

As an example, the following results were obtained for CsCl ($r_e = 0.35706$ nm):

α	A	α	A
2π	1.762620	4π	1.762667
3π	1.762201	5π	1.762663

We draw attention to the fact that different authors calculate or report the Madelung constant in differing ways, leading to apparently discrepant values for A. However, it is necessary to consider how the calculation has been carried out: with respect to a distance r; with respect to a cell side a, b or c; with the ionic charges as unity or with their formal values. We illustrate these differences with the zinc blende structure.

Appendix 2

For β-ZnS, we have a face-centred cubic unit cell of side 0.5409 nm, containing four ZnS entities. The atoms have the fractional coordinates as follow:

$$Zn \qquad (0, 0, 0) + F$$
$$S \qquad (¼, ¼, ¼) + F$$

It follows that r_e = 0.23422 nm; the charges q_+ and q_- are +2 and –2, respectively. Using (A2.1) and the program *MADC*, we obtain A = 6.55232. If we employ unit positive and negative charges, A = 1.63808. If A is calculated with respect to a instead of r_e, the two values become 15.13193 and 3.78298, respectively. It is evident that 6.55232/0.23422 = 1.63808 × 2 × 2/0.23422 = 15.13193/0.5409 = 3.78298 ×2 × 2/0.5409 = 27.975. The advantages of (A2.1) are that the charges are included and that the distance r used in the term A/r is that value used in the calculation of electrostatic energy, whether or no it is a unique value in the structure.

Appendix 3 Reciprocal space

A3.1 INTRODUCTION
Reciprocal space is a device used in solid state studies, mainly by crystallographers and spectroscopists, to interpret experimental results. Although applied in different ways and with differing notations in the solid state, one and the same concept lies behind each such application. We introduce the reciprocal lattice first by means of a geometrical construction.

A3.2 GEOMETRY OF THE RECIPROCAL LATTICE
For each Bravais (direct) lattice in *real* space, a corresponding **reciprocal lattice** may be postulated in *reciprocal* space. It has the *same symmetry* as the direct lattice, and may be derived from it by the following construction. Let Figure A3.1 represent a monoclinic lattice in projection on to the x,z plane. From an origin O, of the P lattice unit cell of delineating vectors *a*, *b* and *c*, normals are drawn to **families** of parallel, equidistant planes (*hkl*) in real space, where h, k and l are the **Miller indices** of the families of planes. The Miller indices h, k and l are the reciprocals of the fractional intercepts made by the plane in the family nearest to the origin, on *a*, *b* and *c*, respectively. We remark *en passant* that the normals employed here do not, normally, coincide with the *directions* (see Section 1.4.5) of the same indices [*hkl*].

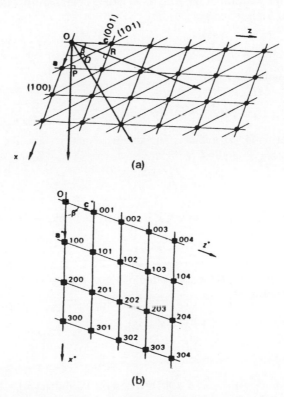

Fig. A3.1. Direct and reciprocal lattices. (a) Monoclinic P lattice, as seen in projection along *b*, showing three families of planes. (b) Corresponding reciprocal lattice, showing the points that represent these three and other families of planes.

The interplanar spacings in real space are denoted by $d(hkl)$. Along each normal, reciprocal lattice points hkl (no parentheses) are marked off such that the distance from the common origin O to the reciprocal lattice point hkl is given by

(A3.1) $$d^*(hkl) = K/d^*(hkl) ,$$

where K is a constant which has the value unity in most spectroscopic work, but takes the value of the wavelength of an X-radiation in most crystallographic applications. We shall take the value of K as unity in this discussion, unless otherwise noted.

The reciprocal lattice unit cell is defined by the three vectors $d^*(100)$, or a^*, $d^*(010)$, or b^*, and $d^*(001)$, or c^*. Where the lattice is not orthogonal, the interaxial angles are defined as $\alpha^* = b^*{}^{\wedge}c^*$, $\beta^* = c^*{}^{\wedge}a^*$ and $\gamma^* = a^*{}^{\wedge}b^*$.

The reciprocal lattice points form a true lattice, as we shall prove in the next section. A reciprocal lattice row hkl; $2h,2k,2l$; $3h,3k,3l$; $\cdots$ may be considered to be derived from families of planes (nh,nk,nl), where $n = 1, 2, 3, \cdots$, since $d(nh,nk,nl) = d(hkl)/n$. It follows that

(A3.2) $$d^*(nh,nk,nl) = nd^*(hkl) ,$$

where $d^*(hkl)$ is the distance of the reciprocal lattice point hkl from the origin. The vector from the origin to hkl is given by

(A3.2) $$d^*(hkl) = ha^* + kb^* + lc^* .$$

It follows from Figure A3.1 that

(A3.3) $$a^* = 1/d(100) = 1/[a \sin(\beta)] ,$$

with similar equation for b^* and c^*. The angles α^* and γ^* here are each 90°, whereas $\beta^* = 180° - \beta$. More generally,

(A3.4) $$a^* = (b \times c)/V ,$$

with cyclic permutations for b^* and c^*; V is the unit cell volume, given by

(A3.5) $$V = a \cdot b \times c ,$$

or its cyclic permutations. The scalar value of V has been given by (1.5), and it has been shown elsewhere[5] that

$$V^* = 1/V.$$

We see that

(A3.6) $$a \cdot a^* = aa^* \cos(\beta - 90) = a \cos(\beta - 90)/[a \sin(\beta)] = 1 ,$$

and generally,

(A3.7) $$a \cdot a^* = b \cdot b^* = c \cdot c^* = 1 ,$$

whereas

(A3.8) $a \cdot b^* = a \cdot c^* = a^* \cdot b$, etc $= 0$.

General formulae exist for the angles, such that

(A3.9) $\cos(\alpha^*) = [\cos(\beta) \cos(\gamma) - \cos(\alpha)] / [\sin(\beta) \sin(\gamma)]$,

with cyclic permutations for β^* and γ^*.

A3.2.1 Truth of the reciprocal lattice
We show that the reciprocal lattice is a true lattice. For any plane (hkl), we have from (A3.1) with $K = 1$

(A3.10) $d(hkl) \cdot d^*(hkl) = 1$.

We know also that

(A3.11) $d(hkl)/(a/h) = \cos(\chi)$,

where χ is the direction cosine of $d(hkl)$ with respect to the x axis. Also

(A3.12) $d(hkl) \cdot a/h = [d(hkl) a/h] \cos(\chi) = d^2(hkl)$.

Using (A3.1), and because $d(hkl)$ and $d^*(hkl)$ are collinear,

(A3.13) $d(hkl) = d^*(hkl)/ d^2(hkl)$.

Hence,

(A3.14) $[d^*(hkl)/ d^{*2}(hkl)] \cdot a/h = d(hkl) \cdot a/h = 1/ d^{*2}(hkl)$,

or

(A3.15) $d^*(hkl) \cdot a = h$.

Similarly,

(A3.16) $d^*(hkl) \cdot b = k$,

(A3.17) $d^*(hkl) \cdot c = l$,

whence

(A3.18) $d^*(hkl) = ha^* + kb^* + lc^*$,

so that the reciprocal lattice points hkl form a true lattice, and any point hkl in reciprocal space represents a family of planes (hkl) in real space.

A3.2.2 Reciprocity of F and I unit cells
In Figure A3.2, we show an F unit cell, and a P unit cell formed from it by the real-space transformation

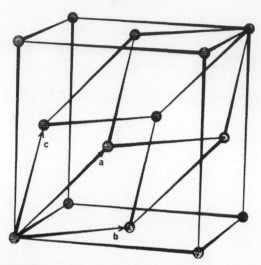

Fig. A3.2. An F lattice unit cell, with the related P unit cell outlined within it. The P unit cell has one quarter of the volume of the F unit cell.

$$a_P = b_F/2 + c_F/2 \ ,$$

(A3.19) $$b_P = c_F/2 + a_F/2 \ ,$$

$$c_P = a_F/2 + b_F/2 \ .$$

Using (A3.4),

(A3.20) $$a_P{}^* = (b_P \times c_P)/V_P = [(\ c_F/2 + a_F/2) \times (a_F/2 + b_F/2)] / V_P \ .$$

Since $V_F = 4V_P$, and because the vector product $a \times a = 0$, this equation may be re-written as

(A3.21) $$a_P{}^* = [(\ c_F \times a_F) + (c_F \times a_F) + (a_F \times b_F)] / V_F \ .$$

Hence,

(A3.22) $$a_P{}^* = -a_F{}^* + b_F{}^* + c_F{}^* \ ,$$

where the negative sign preserves the right-handed nature of the axes. Similar equations can be deduced for $b_P{}^*$ and $c_P{}^*$. Considering next the I unit cell, the equation similar to (A3.22) is

(A3.23) $$a_P{}^* = -a_I{}^*/2 + b_I{}^*/2 + c_I{}^*/2,$$

with similar equations for $b_P{}^*$ and $c_P{}^*$. By writing (A3.22) as

(A3.24) $$a_P{}^* = -2a_F{}^*/2 + 2b_F{}^*/2 + c_F{}^* /2,$$

we see that an F unit cell in a Bravais lattice reciprocates to an I unit cell in reciprocal space, where the I unit cell is defined by the vectors $2a_F{}^*$, $2b_F{}^*$ and $2c_F{}^*$.

A3.3 LINKING THE SOLID STATE APPLICATIONS

The reciprocal space concept as discussed above would be readily applicable in crystallographic work. We need to consider how it is used in other solid state studies, and where the two approaches meet in practice.

In Chapter 4, we discussed the role of k-space in discussing the wave-mechanical theory of the metal bond, and here we shall limit our discussion to cubic crystals. We defined k as a *wave vector* in reciprocal space, and identified its components along the x, y and z axes (see Section 4.3). It is common practice in spectroscopic work to take the constant K in (A3.1) as 2π, so that the distance of the first reciprocal lattice point along the reciprocal axis a^*, or k_x, is $2\pi/a$. The equation

$$(A3.25) \qquad\qquad k = n_x k_x + n_y k_y + n_z k_z$$

is analogous to (A3.18), and defines a distance in k-(reciprocal)-space. Many crystal structures have centred structure unit cells, but it is desirable in spectroscopy to work with primitive unit cells. As we have seen, it is always possible to define a primitive unit cell in any crystal system, but such a cell may not of itself display the symmetry of the lattice. A cell that is both of primitive-type, in that it has one lattice point per unit volume, and displays the full lattice symmetry is the Wigner-Seitz cell, and we digress for a moment to consider this important cell.

We have illustrated already (Figure 4.6, Section 4.4) the Wigner-Seitz cell obtained from a body-centred cubic lattice. It shows six cube faces, the form <100>, and eight octahedron faces <111>; a form of planes <*hkl*> comprises those planes (*hkl*) related by the point-group symmetry of the given crystal.

The cell was obtained by drawing planes to bisect perpendicularly the lines drawn in real space from a given body-centred lattice point to the eight corners of its unit cell and to the body-centred points in the six adjacent unit cells. The mutual intersections formed by these planes define the Wigner-Seitz cell of the I cubic lattice. *No other* bisecting planes can be drawn that would intersect the closed figure.

When the Wigner-Seitz cell is constructed in the reciprocal lattice, it constitutes the *first Brillouin zone* (see Section 4.3ff). In crystallographic practice, the size of the reciprocal lattice is determined by the wavelength of the exploring radiation. A typical X-ray wavelength is *ca* 0.15 nm, so that a cube of side 0.1 nm would have a reciprocal lattice unit cell side of 1.5 dimensionless reciprocal lattice units (*RLU*). In contrast, a spectroscopic examination of the same crystal may be using a wavelength of, say 1000 nm to 5000 nm, so that the reciprocal lattice is then vanishingly small. Effectively, the spectroscopist is examining reciprocal space at and around the value $|k| = 0$; he is not concerned with any of the reciprocal lattice points outside the limits of the first Brillouin zone. On the other hand, the crystallographer uses all available reciprocal lattice points *outside* the first Brillouin zone. In some cases, a shorter wavelength than 0.15 nm is used, so as to increase the size of the reciprocal lattice available for experimental exploration. We may say that the first Brilllouin zone represents a meeting point for these two solid state applications. Further studies of this topic may be found in the standard literature[5,16,20-23].

CHECKLIST A

At the end of these three appendices, you should be able to:

1. Recognize chemical examples of point-group symmetries;
2. Carry out the calculation of Madelung constants;
3. Understand the relationship between the k-space used in spectroscopy and the reciprocal space used in crystallography,

Listed references

[1]   H Pollaczek-Geiringer, *Zeitschrift. für. angewandte. Mathematik und Mechanik* **6**, 70 (1926).

[2]   M F C Ladd, *Symmetry in Molecules and Crystals* (Ellis Horwood, 1989, 1992).

[3]   Mark Ladd, *Symmetry and Group Theory in Chemistry* (Horwood Publishing Limited, 1998).

[4]   International Union of Crystallography, *International Tables for Crystallography*, Volume A (Reidel, 1983); formerly, *International Tables for X-Ray Crystallography*, Volume I (Kynoch Press, 1985).

[5]   M F C Ladd and R A Palmer, *Structure Determination by X-ray Crystallography*, 3rd edition (Plenum Press, 1994).

[6]   G Burns and A M Glazer, *Space Groups for Solid State Scientists* (Academic Press, 1978).

[7]   Mark Ladd, *Chemical Bonding in Solids and Fluids* (Ellis Horwood, 1994).

[8]   A L Allred, *Journal of Inorganic and Nuclear Chemistry* **17**, 215 (1961).

[9]   M P Tosi, *Solid State Physics* **16** (Academic Press, 1964).

[10]   M F C Ladd and W H Lee, *Progress in Solid State Chemistry*, **1** (Pergamon, 1964); *idem. ibid.* **2** (Pergamon, 1965); *idem. ibid.* (Pergamon, 1966).

[11]   D H Templeton, *Journal of Chemical Physics* **23**, 1629 (1955).

[12]   Internet *www.horwood.net/publish*.

[13]   M F C Ladd, *Theoretica Chimica Acta* **12**, 333 (1968).

[14]   H Witte and H Wölfel, *Zeitschrift für physikalische Chemie* (Frankfurt) **3**, 296 (1955); *idem. Review of Modern Physics* **30**, 51 (1958)

[15]   R D Shannon and C T Prewitt, *Acta Crystallographica* B**25**, 925 (1969); *idem, ibid.* B**26**, 1046 (1970)

[16]   Mark Ladd, *Introduction to Physical Chemistry*, 3rd Edition (Cambridge University Press, 1998)

[17]   P W Atkins, *Physical Chemistry*, 5th Edition (Oxford University Press, 1997)

[18]   L Pauling and E B Wilson, *Quantum Mechanics* (McGraw-Hill, 1935).

[19]   R McWeeny, *Symmetry*, (Pergamon, 1963).

[20]   E Wigner and F Seitz, *Physical Review*, **43**, 804 (1933).

[21]   C Kittel, *Introduction to Solid State Physics*, 5th Edition (Wiley, 1976).

[22]   S F A Kettle and L J Norrby, *Journal of Chemical Education*, **71**, 1003 (1994).

[23]   M F C Ladd, *Journal of Chemical Education*, **74**, 461(1997).

[24]   A H Cottrell, *An Introduction to Metallurgy* (Arnold, 1975)

[25]   W Hume-Rothery , R E Smallman and C W Haworth, *The Structure of Metals* (Institute of Metals, 1969).

[26]   F D Rossini *et alia*, *Circular No. 500, and later Supplements* (National Bureau of Standards, 1952 and subsequently).

[27]   J N Israelachvili, *Intermolecular and Surface Forces* (Academic Press, 1992).

[28]   E A Moelwyn-Hughes *Physical Chemistry* (Pergamon Press, 1961).

[29]   A I Kitaigorodskii, *Organic Chemical Crystallography* (Consultants Bureau, 1961).

[30]   M F C Ladd, *Journal of the Chemical Society*, Dalton Transactions, 220 (1977).

[31]   J N Bradley and P D Greene, *Transactions of the Faraday Society*, **63**, 2516 (1967).

[32]   M F C Ladd and W H Lee, *Zeitschrift für Kristallographie*, **129**, 157 (1969).

[33]   Cambridge Crystallographic Data Base, 12 Union Road, Cambridge   CB2 1EZ
(e-mail: teched@chemcrys.cam.ac.uk).

[34]   *Strukturbericht* Volumes 1–7 (1913–1939) continuing as
Structure Reports Volume 8 (1940), through Volume 57A (1990), to date.

[35]   R W G Wyckoff, *Crystal Structures*, Volume 1 (Academic Press, 1963) and
Volumes 2–6 in the same series.

[36]   Editor L E Sutton, *Tables of Interatomic Distances and Configuration in
Molecules and Ions* (The Chemical Society, (1958) and Supplement (1965).

[37]   S F A Kettle and L J Norrby, *Journal of Chemical Education* **67**, 1022 (1990),
and references therein.

[38]   Editor K Hermann, *Internationale Tabellen zür Bestimmung von
Kristallstrukturen*, Volumes I and II (Borntraeger, Berlin, 1935)

[39]   F T Whittaker and G Robinson, *The Calculus of Observations* (Blackie, 1949).

[40]   H M Evjen, *Phys Rev*, **39**, 675 (1932).

Tutorial solutions

SOLUTIONS 1

1.1. Formally, we could write 422, $42\bar{2}$, $4\bar{2}2$, $4\bar{2}\,\bar{2}$, $\bar{4}22$, $\bar{4}2\bar{2}$, $\bar{4}\,\bar{2}2$, $\bar{4}\,\bar{2}\,\bar{2}$. Because the interaction of two roto-inversions leads to a pure rotation, all symbols with one or three roto-inversion axes are not permitted. Now, $\bar{4}2\bar{2}$ and $\bar{4}\,\bar{2}2$ are equivalent under rotation in the x,y plane by 45°, so that we have 422, $42\bar{2}\,\bar{2}$ and $\bar{4}2\bar{2}$. Conventionally, they are written as 422, $4mm$, and $\bar{4}2m$.

1.2. H_2PO_4, 3; C_6H_5Cl, $mm2$; $ThBr_4$, $\bar{4}2m$.

1.3. (i) Orthorhombic $B \equiv$ orthorhombic C, by the transformation
$$\boldsymbol{a}' = \boldsymbol{a}, \quad \boldsymbol{b}' = \boldsymbol{c}, \quad \boldsymbol{c}' = -\boldsymbol{b}.$$

(ii) 'Tetragonal' A is not a tetragonal lattice unit cell, but it could be described as orthorhombic A ($\equiv$ orthorhombic C).

(iii) Triclinic $I \equiv$ triclinic P by the transformation
$$\boldsymbol{a}' = -\boldsymbol{a}/2 + \boldsymbol{b}/2 + \boldsymbol{c}/2, \quad \boldsymbol{b}' = \boldsymbol{a}/2 - \boldsymbol{b}/2 + \boldsymbol{c}/2, \quad \boldsymbol{c}' = \boldsymbol{a}/2 + \boldsymbol{b}/2 - \boldsymbol{c}/2.$$

1.4. Monoclinic $F \rightarrow$ monoclinic C by the transformation
$$\boldsymbol{a}' = \boldsymbol{a}, \quad \boldsymbol{b}' = \boldsymbol{b}, \quad \boldsymbol{c}' = -\boldsymbol{a}/2 + \boldsymbol{c}/2;$$
$c' = 0.5763$ nm, $\beta = 139.47°$; $V_C = V_F/2$ (count lattice points per unit cell).

1.5. 2.874 nm (F); 2.864 nm (C).

1.6. It is not an eighth system, because the symmetry at each lattice point is not higher than $\bar{1}$. It is a special case of the triclinic system with $\gamma = 90°$.

1.7. (i) The symmetry is no longer tetragonal, but the unit cell represents a lattice (orthorhombic).

(ii) The tetragonal symmetry is apparently restored, but the unit cell no longer represents a lattice, because the lattice points do not all have an identical environment.

(iii) A tetragonal F unit cell is obtained, which is equivalent to tetragonal I under the transformation
$$\boldsymbol{a}' = \boldsymbol{a}/2 + \boldsymbol{b}/2, \quad \boldsymbol{b}' = -\boldsymbol{a}/2 + \boldsymbol{b}/2, \quad \boldsymbol{c}' = \boldsymbol{c}.$$

1.8. $P2/c$: monoclinic, $2/m$; P unit cell; $2 // y, c \perp z$.

$Pca2_1$: orthorhombic, $mm2$, P unit cell; $c \perp x, a \perp y, 2_1 // z$.

$Cmcm$: orthorhombic, mmm; $m \perp x, c \perp y, m \perp z$.

$P\bar{4}2_1c$: tetragonal, $\bar{4}2m$; $\bar{4} // z, 2_1 // x$ and $y, c \perp [110]$ and $[\bar{1}10]$.

1.9. Monoclinic; 2_1 along the line $x = 0, y, z = \frac{1}{4}$; c parallel to the plane $x, y = \frac{1}{4}, z$.

1.10. (a) Refer to Figure S1.1.

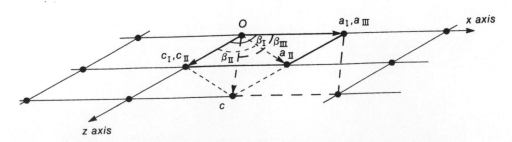

Fig. S1.1. Monoclinic unit cells (i), (ii) and (iii), in projection on the x,z plane.

(b) Cell (i), C; cell (ii), I; cell (iii), F. Cell (iii) would not be chosen as it has twice the volume of cells (i) and (ii). Cell (i) (C) is the conventional choice. Cell (ii) might be justified because it has the same volume as cell (i), but has a less obtuse β-angle, which is sometimes convenient.

1.11. (a) Refer to Figure S1.2. (b) $a_R = a_C/\sqrt{2}$. (c) $\cos(\alpha_R) = a_R \cdot b_R = (b_C/2 + c_C/2) \cdot (c_C/2 + a_C/2)/a_R^2 = c_C^2/4 \times 2/a_C^2 = 1/2$, since $a_C = c$; hence, $\alpha_R = 60°$.

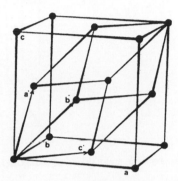

Fig. S1.2. A cubic F unit cell containing a rhombohedral R unit cell.

1.12. (a), (b) Refer to Figure S1.3. (c) The symmetry of 1,2-difluorobenzene is *mm*2. Four molecules would occupy the general equivalent positions d. Two molecules may be placed (i) on positions c, with a space group m-plane coinciding with either the molecular plane or the plane bisecting the C_1—C_2 and C_4—C_5 bonds, or (ii) on positions b or a, with a space group two-fold axis coinciding with the two-fold axis bisecting the C_1—C_2 and C_4—C_5 bonds.

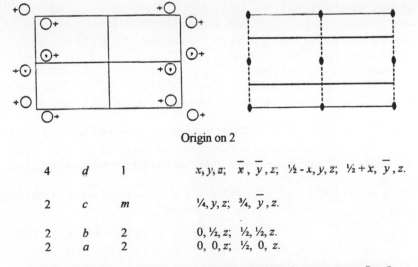

Origin on 2

4	d	1	x, y, z; $\bar{x}, \bar{y}, z$; $\frac{1}{2} - x, y, z$; $\frac{1}{2} + x, \bar{y}, z$.
2	c	m	$\frac{1}{4}, y, z$; $\frac{3}{4}, \bar{y}, z$.
2	b	2	$0, \frac{1}{2}, z$; $\frac{1}{2}, \frac{1}{2}, z$.
2	a	2	$0, 0, z$; $\frac{1}{2}, 0, z$.

Fig. S1.3. Equivalent positions and symmetry elements in space group $Pma2$.

1.13. Refer to Figure S1.4, which is a stereoview of two adjacent F tetragonal unit cells. The unit cell shown by thin lines is I and has the conventional tetragonal conditions, namely, $a = b \neq c$, $\alpha = \beta = \gamma = 90°$. Thus, $F \equiv I$ in the tetragonal system.

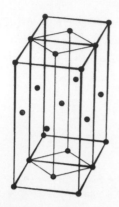 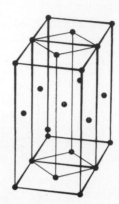

Fig. S1.4. Stereoview to show the equivalence of tetragonal F and tetragonal I.

SOLUTIONS 2

2.1. Following Evjen's procedure, but in two dimensions, we obtain:
(a) 3×3: Take the central charge as an origin, and let the nearest-neighbour distance be unity. Then the terms, with their contribution to the 3×3 cell are: $-(4/1 \times 1/2) + (4/\sqrt{2} \times 1/4) = -1.2928$, so that $A = 1.2928$. Similarly, (b) 5×5: 1.6068, and (c) 7×7: 1.6105. The variation is clearly non-linear. Extrapolation of the results (a) and (b) against 1/(cell side) to zero gives 1.6179, which must be too large. A compromise is (1.6179 + 1.6105)/2, or 1.6142. (The true value, to six significant figures, is 1.61398.)

2.2. For the electrostatic self-energy, we need the sum of all terms such as (2.1) over the ion, that is $e^2/(4\pi\varepsilon_0)[6/(0.2123 \times 10^{-9}\ nm) - 4/(0.130 \times 10^{-9}\ nm)] = -7.68 \times 10^{-18}$ J.

2.3. From the sum of the ionic radii, with a departure of 0.001 nm), $r_e = 0.397$ nm. If the cube side is a, then $2re/\sqrt{3} = a = 0.4584$ nm. Hence, density = $[(132.9 + 126.9) \times 1.6605 \times 10^{-27}\ kg]/(0.4584 \times 10^{-9})^3 = 4479$ kg m^{-3}

2.4. The unique distances are between Ti ($\frac{1}{2}$, $\frac{1}{2}$. $\frac{1}{2}$) and O_1 (x, x, 0) and O_2 ($\frac{1}{2} + x$, $\frac{1}{2} - x$, $\frac{1}{2}$). It may help to sketch the structure unit cell. Hence, Ti– O_1 = 0.1985 nm, Ti–O2 = 0.1945 nm. (The ionic radii were given to enable the correct Ti and O pairs to be chosen.)

2.5. U(MgCl, cr) = -864 kJ mol^{-1}; U(MgCl$_2$, cr) = -2499 kJ mol^{-1}. Although considerably more energy is needed to create Mg^{2+} (g) compared with Mg$^+$ (g), it is more than balanced by the lower lattice energy of MgCl$_2$ (cr) compared with MgCl (cr), so that the former compound is preferred.

2.6. U(CaO, cr) = $-(10^3 L\ mol^{-1}) \times 1.74756 \times 4 \times e^2(1 - 0.162)/(4\pi\varepsilon_0 \times 0.24055 \times 10^{-9}$ nm). Thus, U(CaO, cr) = -3383; hence, E(O^{2-}) = 609 kJ mol^{-1}.

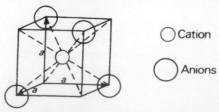

Fig. S2.1. Coordination pattern in würtzite (and zinc blende).

2.7. The essence of the würtzite structure is, like zinc blende, a zinc atom coordinated by four sulfur atoms. From Figure S2.1, when the atoms are in maximum contact, $a\sqrt{2} = 2r_-$ and $a\sqrt{3} = 2r_+ + 2r_-$. Hence, $R = r_+/r_- = 0.225$.

2.8. A(rutile), using $r_e = 0.1945$ nm, is 19.07909. A(anatase), using $r_e = 0.1910$ nm, is 18.7953. Electrostatic energy is proportional to A/r_e; thus U(rutile) $\propto -98.09$, and U(anatase) $\propto -98.40$. This result indicates that anatase is likely to be more stable than rutile, as discussed in the text. However, one cannot always deduce the relative stabilities of compounds from their electrostatic energies alone.

2.9. (a) The electrostatic bond strength Be—H_2O is 1/2; thus, H_2O—SO_4 is 1/4, because the water molecule is neutral. Each pair of O_3SO—$(H_2O)_2$ links provide $2 \times 1/4$; there are eight of them in all, which sum to the charge on the SO_4 group. Thus, the electrostatic valence rule is obeyed in this hydrate. (b) A better formulation would be Be$(H_2O)_4$ SO_4. (c) The water molecules distribute the charge on beryllium over a larger region than that of the bare cation, thereby reducing its polarizing effect.

2.10. Let O_1 and O_2 have the coordinates (x_1, y_1, z_1) and (x_2, y_2, z_2), respectively. Using a relation like (1.8), we obtain $r_{12} = [(x_2 - x_1)^2 a^2 + (y_2 - y_1)^2 b^2 + (z_2 - z_1)^2 c^2 + 2(x_2 - x_1)(z_2 - z_1)ac \cos(\beta)]^{\frac{1}{2}}$. Hence, $O_1 \ldots O_2 = 0.246$ nm, $O_1 \ldots H = 0.120$ nm, $O_2 \ldots H = 0129$ nm. If the coordinates of the H atom are (x_0, y_0, z_0), application of (1.9) leads to $r_{01} \cdot r_{02} = r_{01} r_{02} \cos(\theta)$, where θ is the angle $O_1 \ldots H \ldots O_2$, and r_{01} and r_{02} are obtained as above. In extenso, $r_{01} \cdot r_{02} = (x_0 - x_1)(x_0 - x_2)a^2 + (y_0 - y_1)(y_0 - y_2)b^2 + (z_0 - z_1)(z_0 - z_2)c^2 + [(x_0 - x_1)(z_0 - z_2) + (x_0 - x_2)(z_0 - z_1)ac \cos(\beta) = r_{01}r_{02} \cos(\theta)$, whence $\theta = 163.5°$. The hydrogen bond is unsymmetrical and non-linear.

2.11. $NaNbO_3$: ε(Na—O) $= 1/12$, ε(Nb—O) $= 5/6$; strength at O is $|4/12| + |10/6| = 2$. $KZnF_3$: ε(K—F) $= 1/12$, ε(Zn—F) $= 2/6$; strength at O is $|4/12| + |4/6| = 1$. Fe_3O_4: ε[Fe(III)—O] $= 3/4$ and $3/6$, ε[Fe(II)—O] $= 2/6$; strength at O $= |3/4| + |3/2 \times 3/6| + |3/2 \times 2/6| = 2$. Thus, the electrostatic valence rule is satisfied in each compound. The inverse spinel Fe_3O_4 is better formulated as Fe(III)[Fe(II)Fe(III)]O_4.

2.12. From the limiting law, $\gamma_\pm = 0.70$. Thus, $\Delta G_d° = -RT \ln(0.09 \times 0.70)^2 = 13.7$ kJ mol^{-1}. From the thermodynamic data, $\Delta G_d° = 4.6 - 0.298(4.6 - 36.0) = 14.0$ kJ mol^{-1}. The small discrepancy arises from the approximate nature of the limiting law.

2.13. At 0 K, $(\partial U/\partial V)_T = -p$, so that $(\partial^2 U/\partial V^2)_T = -(1/V)(\partial p/\partial V)_T$. At constant temperature, $d^2 U/dV^2 = 1/(\kappa V)$. For the NaCl structure type, we can write $V = cr^3$, where c is a constant. Hence, $dV/dr = 3cr^2 = 3V/r$, and $d^2 V/dr^2 = 6cr = 6V/r^2$. From the rules for differentiation, $dU/dV = dU/dr \div dV/dr$ and $d^2 U/dV^2 = 1/(dV/dr) d/dr [dU/dr \div dV/dr]$. Then, $d^2 U/dV^2 = 1/(dV/dr)^3 [(dV/dr)(d^2 U/dr^2) - (dU/dr)(d^2 V/dr^2)]$. From (2.6), $(d^2 U/dr^2) = 2A'/r^3 + a^2 B \exp(ar)$. Hence, $1/(\kappa V) = 1/(3V/r)^3 \{(3V/r)[2A'/r^3 + a^2 B \exp(ar)] - (6V/r^2)[-A'/r + aB \exp(ar)]\}$. Eliminating the constant B from (2.7), for $r = r_e$, we find $9V/\kappa = 2A'/r_e + A'a = 2A'/r_e + (A'/r_e)ar_e$. Introducing the true values for V, A' and a and rearranging, remembering that $q_+q_- = -1$ for NaCl, gives $p/r_e = Ae^2/(4\pi\varepsilon_0 r_e) = Ae^2/(4\pi\varepsilon_0 r_e)/[18r^3/\kappa + 2Ae^2/(4\pi\varepsilon_0 r_e)]$.

(b) For NaCl, $A = 1.7476$, so that $p/r_e = 1.4298 \times 10^{-18}/[(9.8456 \times 10^{-18}) + (2.8596 \times 10^{-18})] = 0.113$. Thus, the lattice energy for sodium chloride, from (2.8) is -764 kJ mol^{-1}.

2.14. The ionization energy and electron affinity refer to the processes

$$M(g) \to M^+(g) + e,$$

and

$$X(g) + e \to X(g).$$

The first reaction at T adds $5RT/2$ to $\Delta H_i°$, while the second reaction subtracts $5RT/2$ from $\Delta H_e°$, so that the two quantities cancel in the cycle.

SOLUTIONS 3

3.1. (a) Since momentum $p = mv$, $\delta p = m\delta v$, so that $\delta v = (\hbar/2)/(m\delta x) = 1.05 \times 10^{-28}$ m s^{-1}.
(b) $\delta v = 57.9$ m s^{-1}.

3.2. $N^2 \int_0^\infty \exp(-2r/a_0) \, r^2 \, dr \int_0^\pi \sin(\theta) \, d\theta \int_0^{2\pi} d\phi = 1 = 4\pi \, N^2 \int_0^\infty \exp(-2r/a_0) \, r^2 \, dr.$

Generally, by the reduction formula, $\int r^n \exp(cr) \, dr = (x^n/c) \exp(cr) - (n/c) \int r^{n-1}\exp(cr)$
dr. Thus, putting $c = -2/a_0$, we obtain $1 = 4\pi N^2 \times 2a_0^3/8$, so that $N = (\pi a_0^3)^{-\frac{1}{2}}$.

3.3. The box length a is 1.12 nm, and there are eight electrons, two from each double
bond and two from the neutral nitrogen atom. The first four energy levels are filled by
two spin-paired electrons, so that the lowest-energy transition is from E4 to E5. Thus, ΔE
$= E_5 - E_4 = 9h^2/(8m_e a^2)$, which gives $\Delta E = 4.323 \times 10^{-19}$ J.

3.4. Differentiate the function, and set the derivative to zero. Thus, ignoring constants,
$r2(-2/a_0) \exp(-2r/a_0) + \exp(-2r/a_0) (2r) = 0$, so that the maximum occurs at $r = a_0$.

3.5. (a) C: $(1s)^2 (2s)^2 (2p_x)^1 (2p_y)^1$; (b) O: $(1s)^2 (2s)^2 (2p_x)^2 (2p_y)^1 (2p_z)^1$; (c) F: $(1s)^2 (2s)^2$
$(2p)^6$.

3.6. (a) σ_u; (b) σ_g; (c) $\sigma_u + \pi_u$.

3.7. If the bond angle is ϕ, then $-\cos(\phi) = (1/3)/(2/3)$, so that $\phi = 120°$.

3.8. Let the charge on O be q, so that that on each H is $q/2$. Resolving the bond moments
along a line bisecting the H—O—H angle gives an effective length of 2(0.096)
$\cos(104.4/2)°$ nm. Hence, $q = (1.8 \times 3.3356 \times 10^{-30}$ C m$)/(2 \times 0.096 \times \cos(52.2)° \times 10^{-9}$
m $\times 1.6022 \times 10^{-19}$ C$) = 0.318$. Thus, $q_O = -0.318$ and $q_H = +0.159$.

3.9. The atoms are at (0, 0, 0; ¼, ¼, ¼) +F. Hence, C—C = $(a^2/4 + a^2/4 + a^2/4)^{\frac{1}{2}} =$
0.1545 nm.

3.10. The equation for a bond between atoms i and j in the hexagonal system is:
$r_{ij} = \{[(x_i - x_j)^2 + (y_i - y_j)^2 - (x_i - x_j)(y_i - y_j)]a^2 + (z_i - z_j)^2c^2 + 2(x_i - x_j)(y_i - y_j)ab \cos(\gamma)\}^{\frac{1}{2}}$
and it is easily programmed (or see Section 6.2.7). The following results are selected:
Si(0.465, 0, 0)—O(0.415, 0.272, 0.120)   0.1611 nm;
O(0.143, -0.272, -0.120)—O(0.415, 0.272, 0.120)   0.2653 nm;
O(0.272, 0.415, 0.547)—O(0.415, 0.272, 0.120)   0.2609 nm.

SOLUTIONS 4

4.1. For a sphere of radius r, volume = $4\pi r^3/3 = 4.189 \, r^3$. From Figure S4.1, $c = 4r \sin$
$[\cos^{-1}(1/\sqrt{3}) = 3.266r$. Volume of unit cell containing *one* sphere = $\frac{1}{2}a2c \sin(120)°$.
Since $a = 2r$, this volume is $5.657r^3$, so that packing fraction is 0.74.

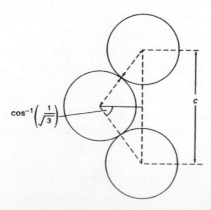

$\cos^{-1}\left(\dfrac{1}{\sqrt{3}}\right)$

Fig. S4.1. Construction for the unit cell dimension c in the close-packed hexagonal structure type.

4.2. $E_F = [h^2/(8\pi^2 m_e)][3\pi^2(N/V)]^{2/3}$; but $N/V = 2/a^3$, where $a = 0.350$ nm, so that $E_F = 7.57 \times 10^{-19}$ J, or 4.73 eV. Average electronic energy is $3E_F/5$, which may be equated to $3k_B T_F/2$, so that $T_F = 2.19 \times 10^4$ K.

4.3. (a) See Figure S4.2; (b) A cube of side $2\pi/a$.

4.4. The first Brillouin zone is a cube of side $2\pi/a$. Hence, the sphere radius k_F is equal to π/a. Using $N/V = k_F^3/3\pi^2 = Zn_e/a3$, we have $Zn_e/a^3 = \pi/(3a^3)$, so that $n_e = \pi/(3Z)$.

4.5. Eight tetrahedral holes: ($\frac{1}{4}$, $\frac{1}{4}$, $\frac{1}{4}$; $\frac{1}{4}$, $\frac{1}{4}$, $\frac{3}{4}$)$+ F$; four octahedral holes: ($\frac{1}{2}$, 0, 0)$+ F$.

4.6. Using $E = hc/\lambda$, $\lambda = 512$ nm. From $\sigma = A \exp[-\Delta E/(2k_B T)]$, $\rho = 1/\sigma \propto$ $\exp[\Delta E/(2k_B T)]$. $\bar{r}$ Thus, $\rho_{300}/\rho_{400} = \exp[(\Delta E/k_B)(1/600 - 1/800) = 1.21 \times 10^5$.

4.7. Using $R = A' \exp(\Delta E/2k_B T)$, a plot of $\ln(R/\Omega)$ against $1/(T/\text{K})$ will be linear:

$$\ln(R/\Omega) = \ln(A') + [\Delta E/(2k_B)] [1/(T/\text{K})].$$

The slope is $\Delta E/(2k_B)$ which, by least squares is 4270.3 K, so that $\Delta E = 1.179 \times 10^{-19}$ J, or 0.736 eV.

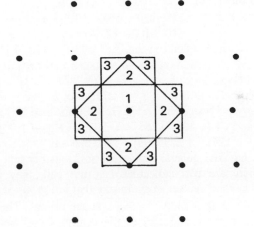

Fig. S4.2. Two-dimensional, square lattice, showing the first three Brillouin zones.

4.8. Following (4.22),

$$\bar{r} = \frac{\int_{-\infty}^{\infty} r \exp[\beta(-ar^2 + br^3)]\, dr}{\int_{-\infty}^{\infty} \exp[\beta(-ar^2 + br^3)]\, dr},$$

where $\beta = 1/(k_B T)$. Expanding:

$$\bar{r} = \frac{\int_{-\infty}^{\infty} [r \exp(-\beta ar^2) + \beta br^4 \exp(-\beta ar^2)]\, dr}{\int_{-\infty}^{\infty} [\exp(-\beta ar^2) + \beta br^3 \exp(-\beta ar^2)]\, dr}.$$

The integrals are, in order, $1/(2\beta a)$, $(3\beta b/8)\sqrt{\pi}(\beta a)^{5/2}$, $(\pi\beta a/4)^{1/2}$ and $\frac{1}{2}\beta b/(\beta a)^2$. Thus, $\bar{r} = 3/(4\beta a)$, or $\frac{3}{4}(k_B T/a)$, so that $\bar{r}$ is directly proportional to T.

4.9. Space group $Fm3m$, with atoms in Wyckoff a positions, or $(0, 0, 0)+ F$.

4.10. From the definition of heat capacity at constant volume, $C_{v,el} = \mathrm{d}E/\mathrm{d}T = 2g(E_F)k_B^2 T$. Thus, the constant b in (4.17) is $2g(E_F)k_B^2$. The total number N of free electrons is given by $\int_0^{E_F} \alpha\sqrt{E}\,\mathrm{d}E$, so that $N = 2\alpha E_F^{3/2}/3$; $g(E_F) = \mathrm{d}N/\mathrm{d}E = \alpha E_F^{1/2}$ which, by substitution for α, becomes $3N E_F^{1/2}/E_F^{3/2} = 3N/E_F$. $E_F(\mathrm{Cu}) = 7.06$ eV $= 7.06e$ J. For 1 mol, $N = L$; hence, $g = 3Lk_B^2/E_F = 3.04 \times 10^{-4}$ J, so that $C_{v,el}$ (25 °C) = 0.091 J mol^{-1}.

4.11. $\bar{6}\,m2$.

4.12. $D/\mathrm{kg\ m^{-3}} = 1 \times 210.0 \times 1.6605 \times 10^{-27}/(0.335 \times 10^{-9})^3$, so that $D = 9275$ kg m^{-3}. The atomic radius is $a/2$, or 0.1675 nm. In a face-centred cubic structure, the face-diagonal is $a\sqrt{2} = 4r = 0.670$ nm. Now, $D/\mathrm{kg\ m^{-3}} = 4 \times 210.0 \times 1.6605 \times 10^{-27}/(0.4738 \times 10^{-9})^3$, so that $D = 13114$ kg m^{-3}.

SOLUTIONS 5

5.1. We use $V(r)/\mathrm{J} = -0.75 \times 1.66^2 \times 15.76 \times 1.6022 \times 10^{-25}/(r/\mathrm{nm})^6 = -5.22 \times 10^{-24}/(r/\mathrm{nm})^6$.
(a) -2.84 J, or -1.71 kJ mol^{-1}; (b) -1.27 J, or -0.77 kJ mol^{-1}.

5.2. $r_e = 2^{1/6}\delta = 0.46$ nm. For $r = r_e$, $V = \varepsilon_{LJ} = 3.16 \times 10^{-21}$ J. $V/\mathrm{J} = 1.265 \times 10^{-20}[(\delta/r)^{12} - (\delta/r)^6]$. (a) $r = 0.52$ nm: $V = -2.31 \times 10^{-21}$ J. (b) $+2.34 \times 10^{-21}$ J.

5.3. For carbon at 0, 0, 0, the bonded oxygen is at 0.110, 0.110, 0.110. Hence, C—O = $0.110a\sqrt{3}$ nm $=0.106$ nm. The intermolecular distance can be calculated between, for example, O (x, x, x) and O $(x, \frac{1}{2} - x, \frac{1}{2} + x)$, giving $[(0.28^2 + 0.5^2)]^{1/2}\,a$ nm = 0.319 nm.

5.4. The transformation to $Pnma$ is $a' = b = 1.2735$ nm, $b' = c = 0.5963$ and $c' = a = 0.4325$ nm. In space group $Pbnm$, the bonded Hg—Cl distance is, for example, between Hg (0.050, 0.126, ¼) and Cl (0.406, 0.255, ¼). Hence, Hg—Cl = $\{[(0.406 - 0.050) \times 0.4325$ nm$]^2 + [(0.255 - 0.126) \times 1.2735$ nm$]^2\}^{1/2} = 0.225$ nm.

5.5. Space group $P2_1/c$ has four general equivalent positions per unit cell (consider point group $2/m$). Since there are two molecules per unit cell, they must lie in special equivalent positions. They are centres of symmetry that are grouped into pairs under the space group, for example, 0, 0, 0 and 0. ½, ½ ($\bar{1}$ at the origin). The planar molecule is also centrosymmetric, and the mid-points of the central C—C bonds lie on a set of centres of symmetry. The crystal structure would be determined if the positions of one phenyl ring were known.

5.6. The central carbon atom lies on the $\bar{4}$ axis, and the coordinates generated by it from x, y, z are $\bar{y}, x, \bar{z}$, $\bar{x}, \bar{y}, z$ and $y, \bar{x}, \bar{z}$. The coordinates for the other molecule can be formed by adding 1 to each x coordinate; there is no need to do this to C(1). Then, calculation gives C(1)—C(2) = 0.150 nm, C(2)—O = 0.146 nm, and the shortest intermolecular distance (between two oxygen atoms) is 0.270 nm. The sum of the van der Waals distance for two oxygen atoms is 0.28 nm, so that we can deduce that weak hydrogen-bonding is present. They join the molecule in layers that are parallel to the x,y plane.

5.7. In dicyanoethyne, π-electron overlap can be brought into play with the packing mode adopted. Evidence for this additional interaction is the intermolecular distance of 0.32 nm, whereas the sum of the van der Waals radii is 0.335 nm. In the hexane structure, the sum of the van der Waals radii is not significantly different from the intermolecular contact distance.

5.8. In the monoclinic system, the unique axis is b. Hence, interchanging a and c in $P2_1/a$ leads to $P2_1/c$, but we change the sign of b (or a or c) to maintain right-handed axes. In

$P2_1/n$, the glide translation is $(a + c)/2$. For this to become $c/2$, a new c dimension, c', is chosen such that $c' = a - c$. The same result is obtained easily by sketching the space group on the x,z plane.

5.9. (a) The energy of interaction is based on two pairwise additive (free from induction) Coulombic terms, Figure S5.1.

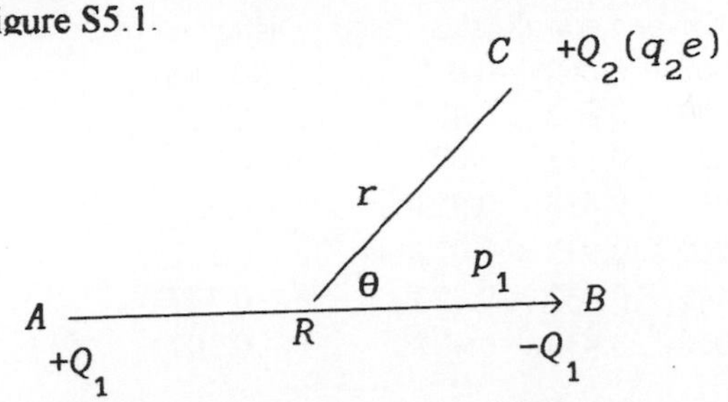

Fig. S5.1. Dipole of moment p_1 distant r from a charge $+Q_2$.

$$V_{i,d} = 1/(4\pi\varepsilon_0)(Q_1Q_2/AC - Q_1Q_2/AC) .$$

$$AC = r[1 + (R/r)\cos(\theta) + R^2/(4r^2)] ; \quad BC = r[1 - (R/r)\cos(\theta) + R^2/(4r^2)] .$$

Hence,

$$V_{i,d} = Q_1Q_2/(4\pi\varepsilon_0 r) \{ [1 + (R/r)\cos(\theta) + R^2/(4r^2)]^{-\frac{1}{2}} - [1 - (R/r)\cos(\theta) + R^2/(4r^2)]^{-\frac{1}{2}} \} .$$

Applying a binomial expansion, to terms no larger than R^2,

$$V_{i,d} = Q_1Q_2/(4\pi\varepsilon_0 r) [R\cos(\theta)] [3R^2/(8r^2) - 1] .$$

Since $R \ll r$ and $p_1 = Q_1R$, we write

$$V_{i,d} = -p_1Q_2 \cos(\theta)/(4\pi\varepsilon_0 r^2) .$$

(b) Since $\theta = \pi$, $V_{i,d} = +(1.5 \times 3.3356 \times 10^{-30}$ C m) $(0.5 \times 1.6022 \times 10^{-19}$ C) / ($4\pi \times 8.8542 \times 10^{-12}$ F m^{-1}) $(0.5 \times 10^{-9}$ nm)$^2 = + 1.44 \times 10^{-20}$ J.

5.10. At $r = r_e$, $\delta/r_e = 2^{-1/6}$, so $V(r_e)/$kJ mol$^{-1} = 6.0221 \times 10^{20} \times 4 \times \varepsilon_{LJ} \times 1.3807 \times 10^{-23} \times 2^{-1/6} (12.132 - 14.454) = 0.06880 \varepsilon_{LJ}$. Hence, for neon, $V(r_e) = 2.46$ kJ mol^{-1}, and for krypton, $V(r_e) = 11.2$ kJ mol^{-1}.

SOLUTIONS 6

6.1. Using the program *LSLI*, the slope of the line from P_m against $1/T$, namely, $Lp^2/(9\varepsilon_0 k_B)$, is 5.273×10^{-2} m^3 mol^{-1}. Hence, $p = 5.273 \times 10^{-30}$ C m, or 1.58 D. Resolving any two bond moments on to the third N—H bond gives a charge of $0.74e$ for the nitrogen atom, with $0.74e/3$ for each hydrogen atom.

6.2. Total number of ions per unit cell = 12; $r_e = (a\sqrt{3})/4 = 0.2366$ nm. Using the program *MADC*, $A(CaF_2) = 5.0388$ (to five significant figures).

6.3. (a) $mm2$; (b) $3m$.

6.4. Using the program *INTP*, $n(25\ °C) = 1.33252$.

6.5. Select a nitrogen atom, say **0.121, 0.244, 0.063**. Generate the four symmetry-related oxygen atoms, and then those related by translation of ±1 along x, y and z separately, 28 oxygen atoms in all, plus the 1 nitrogen atom. Using *BOND*, the following close contacts of the given nitrogen atom were found:

Oxygen coordinates			Distance/nm
0.060	−0.062	−0.023	0.148
0.440	0.062	0.477	0.318
0.560	0.562	0.023	0.350
−0.060	0.438	0.523	0.273
0.060	0.938	−0.023	0.311
0.440	0.062	−0.523	0.377
−0.060	0.438	−0.477	0.307

The distance 0.148 nm corresponds to the N—O bond in the NH_2OH molecule. Of the other intermolecular contact distances, 0.274 nm, 0.307 nm and probably 0.311nm are hydrogen bonds; the sum of the van der Waals radii for N and O is 0.290 nm. The other lengths generated by the program are non-bonded O⋯⋯O distances, some of which may correspond to hydrogen-bonded interactions.

Index

FUNDAMENTALS OF INORGANIC CHEMISTRY
An introductory text for degree course studies
JACK BARRETT, Imperial College of Science, Technology and Medicine, University of London *and* MOUNIR A. MALATI, Mid-Kent College of Higher and Further Education, Chatham
ISBN: 1-898563-38-1   *ca.* 320 pages   1997

Two well-known and experienced teachers offer a foundation course for 1st and 2nd year undergraduate inorganic chemists, covering the main underlying theoretical ideas for an understanding of inorganic chemistry. It takes account of the lower level of mathematical ability among present-day students commencing university study. The necessary mathematics, clearly explained where appropriate, is deliberately non-rigorous.

Undergraduates reading chemistry will find much benefit from these teachers' proper and kindly approach which will launch them into their more advanced part of the inorganic chemistry degree course. The book will be helpful also to those reading any of the sciences where chemistry forms a significant part.

Contents: Introduction; Nuclear and radiochemistry; Electronic configurations and electronic states; Symmetry and group theory; Diatomic molecules and covalent bonding; Polyatomic molecules and metals; Ions in solids and solutions; Chemistry of s-block elements; Chemistry of *p*-block elements; Co-ordination compounds; Chemistry of *d*- and *f*-block elements.

CHEMISTRY IN YOUR ENVIRONMENT: User-friendly, Simplified Science
JACK BARRETT, Department of Chemistry, Imperial College, University of London
ISBN 1-898563-01-2   250 pages   90 diagrams   Hardback   1994
ISBN 1-898563-03-9   Paperback

Introduces chemical "mysteries" and shows the importance of chemistry to life quality, and how physics, metallurgy, geology and engineering, and the sciences of chemistry. biology, biochemistry and microbiology, have contributed to our use of the Earth's resources. Highlights the beneficial and harmful uses of chemicals, and the benefits chemists of made to industry, agriculture, medicine and other human activity.

New Scientist
"An admirable simple course on chemistry. It demands concentration, but it will be rewarded"

Dr John Emsley in *Imperial College Reporter*
"A super book which I thoroughly recommend ... pundits often get things wrong because so many of them are seriously chemically challenged. I suggest you take a large dose of *Chemistry in your Environment*. Dr Barrett's book is so welcome"

SYMMETRY AND GROUP THEORY IN CHEMISTRY
MARK LADD, Department of Chemistry, University of Surrey
ISBN 1-898563-39-X   450 pages   1998

This clear, logical and up-to-date is copiously illustrated, including stereoviews of molecules and hints on stereoviewing. The author's programs aiding the derivation, study and recognition of point groups, and are available on Internet *www.horwood.net/publish*.

Contents: Symmetry everywhere; Symmetry operations and symmetry elements; Group theory and point groups; Representations and character tables; Group theory and wave functions; Group theory and chemical bonding; Group theory, molecular vibrations and electron transitions; Group theory and crystal symmetry.

Education in Chemistry: TheRoyal Society of Chemistry
"A thorough analysis of chemical applications and a valuable advanced treatment for postgraduates and final year undergraduates, providing examples and problems for teaching. Good value and should be bought for the library."

Professor The Lord Lewis, Warden, Robinson College, Cambridge
"Provides the enabling background to rationalize and synthesize the use of symmetry to problems in a wide range of chemical applications, and is a necessary part of any modern course of chemistry."

Dr John Burgess, Reader in Inorganic Chemistry, University of Leicester
"The book treats the subject matter of its title in a logical sequence, thoroughly, and in depth. It would form an excellent text for a major core second or third year course."

IONS IN SOLUTION, Second Edition
JOHN BURGESS, Department of Chemistry, University of Leicester
ISBN: 1-898563-50-0   *ca.* 220 pages   1999

This up-to-date outline of the principles and chemical interactions in inorganic solution chemistry delivers a course module in an important area of considerable complexity which is only cursorily covered in most undergraduate texts.

Problems with solutions and tutorial hints to test comprehension have been added as a feature to check readers' understanding and assist self-study. Exercises and projects are also provided to help readers deepen and extend their knowledge and understanding.

Contents: Introduction; Solvation numbers; Ion-solvent distances; Ion-solvent interactions; Acid-base behaviour: hydrolysis and polymerisation; Stability constants; Redox potentials; Kinetics and thermodynamics; Kinetics and mechanisms: solvent exchange; Kinetics and mechanisms: complex formation; Kinetics and mechanisms: substitution at complexions; Kinetics and mechanisms: redox reactions; Past, present, future predictions.

FUNDAMENTAL CHEMICAL KINETICS

MARGARET ROBSON WRIGHT, School of Chemistry, University of St Andrews
ISBN 1-898563-60-8   *ca.* 320 pages   1999

This innovative and unusual approach delivers for final honours and post graduate students a clear and developmental explation of one of the "harder" areas of physical chemistry, and which is central to all chemistry. Students learn and understand best when the arguments are progressively explained in detail, step-by-step. The text clarifies complex aspects of the more theoretical areas of chemical kinetics, and the significance of the equations which so often baffle students.

Dr.Wright is at great pains to provide detailed verbal clarification of the concepts and of their importance to chemists. Her explanations are an essential and vital feature of the text, after which the mathematical development, like the concepts behind the theory, is then explained in detail to help students to work through on their own. The text is scholarly, lucid and well written, with a combination of depth of coverage and clarity.

Contents: Introduction; Basic Transition State Theory; Advanced Transition State Theory; Basic Unimolecular Reaction Theory; Advanced Unimolecular Reaction Theory; Potential Energy Surfaces; Modified Collision Theory; State to State Kinetics; Energy Transfer; Results of Molecular Beam Experiments.

EXPERIMENTAL INORGANIC/PHYSICAL CHEMISTRY: An Investigative, Integrated Approach to Practical Project Work

MOUNIR A. MALATI, Mid-Kent College of Higher & Further Education, Chatham
ISBN 898563-47-0   *ca.*300 pages   1999

This introduction to instrumental and radiochemical techniques and qualitative and quantitative (volumetric and gravimetric) analysis, with preparation of compounds. It strengthens analytic and preparative skills, and emphasises investigative work. All main elements and groups of the periodic table are covered. Most experiments in the text have been published and proven, and can be carried out using readily available instruments and equipment and inexpensive chemicals. Safety is emphasised.

Contents: Preliminary experiments; Alkali metals; Alkaline earth metals; Boron and aluminium; The carbon group; The nitrogen group; The oxygen group; The halogens; Titanium; Vanadium; Chromium; Manganese; Iron; Cobalt; Nickel; Coinage Metals; The zinc group; Other experiments; Properties of radiation; Neutron activation analysis; Szilard chalmers processes; Exchange reactions; Radiometric titration.

BIOINORGANIC CHEMISTRY, Second Edition
R.F. HAY, Department of Chemistry, University of St Andrews

ISBN: 1-898563-45-4   *ca.* 250 pages   1999

This text provides a general background to the area of biological inorganic chemistry for final year and postgraduate courses and those beginning research in the field. The book has been modernised, restructured, updated and expanded. There is much new material and sections dealing with the functions of different metal ions (nickel, copper, vanadium, chromium, zinc), bio-mineralisation, non-metallic inorganic elements, inorganic drugs and mercury, aluminium, beryllium and chromium (VI).

Journal of Organometallic Chemistry (G.J. Leigh, University of Sussex):
"Probably the best yet produced for undergraduate teaching ... of value for those wishing to start bioinorganic research. I have no doubt that undergraduates and teachers will find it very useful."

REACTIONS MECHANISMS OF METAL COMPLEXES
R.W. HAY, School of Chemistry, University of St. Andrews

ISBN: 1-898563-41-1   200 pages   2000

This text provides a general background as a course module in the area of inorganic reaction mechanisms, suitable for advanced undergraduate and post-graduate study and/or research. The topic has important research applications in the metallurgical industry, and is of interest in the sciences of biochemistry, biology, organic, inorganic and bioinorganic chemistry.

In addition to coverage of substitution reactions in four-, five- and six-coordinate complexes, the book contains further chapters devoted to isomerisation and racemisation reactions, to the general field of redox reactions, and to the reactions of coordinated ligands. This last area is not normally covered in undergraduate texts. It is strongly relevant in other fields such as organic, bioinorganic and biological chemistry, providing a bridge to organic reaction mechanisms. The book also contains a chapter on the kinetic background to the subject with many illustrative examples which will prove very useful to those beginning research.

Contents: Introduction to the field; The kinetic background; Substitution reactions of octahedral complexes; Substitution reactions in four- and five-coordinate complexes; Isomerisation and racemisation reactions; Reactions of coordinated ligands.

ORGANIC CHEMISTRY: A comprehensive degree text and source book
H. BEYER, Institute of Organic Chemistry Ernst-Moritz-Arndt University, Greifswald, Germany *and*
W. WALTER, Institute of Organic Chemistry, University of Hamburg
Translator and Editor: **DOUGLAS LLOYD**, University of St. Andrews, Scotland
ISBN 1-898563-37-3   1038 pages   1997

This book fills a need felt for several decades by providing a standard text from European sources. It provides a background book in support of lecture courses serving for reference and consultation. It will also, because of its vast amount of information, serve for post-graduate use, and a reference source for practising chemists in industry and academia.

Angewendte Chemie International (Germany):
"Its virtue lies in its complementarity to the other sources available in single-volume format ... includes much of industrial relevance, largely missing from modern textbooks, and pays due attention to biochemistry."

Journal of the American Chemical Society:
"A prime example of the European Lehrbuch at it's best: thoroughly up-to-date, comprehensive, lucidly written. To my knowledge there is no comparable book for students with basic knowledge of organic chemistry: it is also a reference book for chemists at any stage of their career."

Chemistry in Britain:
"Contains a considerable amount of up-to-date chemistry, a delightful blend old and new. The significant feature is the vast amount of material it contains. There are sections: general, aliphatics, alicyclics, carbohydrates, aromatics, isoprenoids, hetero-cyclics, amino acids *etc*, nucleic acids, enzymes, and metabolic processes. An invaluable reference, the chemical equivalent of a sedate gentlemen's club."

Chemical Education Today (USA) (John C. Cochran, Department of Chemistry. Colgate University, Hamilton, USA)
"Very appropriate as an intermediate text ... can also be recommended for any organic chemist as a first place to turn for information on a compound or a reaction."

Chemistry in Industry (Society of Chemical Industry Journal):
"Primarily aimed at degree students, it is also a valuable reference for professional chemists in industry or academia."

ANTIOXIDANTS in Science, Technology, Medicine and Nutrition
GERALD SCOTT, Professor Emeritus in Chemistry, Aston University, Birmingham
ISBN 1-898563-31-4   350 pages   1997

The use of antioxidants is widespread throughout the rubber, plastics, food, oil and pharmaceutical industries. This book brings together information generated from research in quite separate fields of biochemical science and technology, and integrates it on the basis of the common mechanisms of peroxidation and antioxidant action. It applies present knowledge of antioxidants to our understanding of their role in preventing and treating common diseases, including cardiovascular disease, cancer, rheumatoid arthritis, ischaemia, pancreatitis, haemochromatosis, kwashiorlor, disorders of prematurity, and diseases of old age. Antioxidants deactivate certain harmful effects of free radicals in the human body due to biological peroxidation, and thus provide protection against cell tissue damage.

Contents: Peroxidation in chemistry and chemical technology; Biological effects of peroxi-dation; Chain-breaking antioxidants; Preventive antioxidants and synergism; Antioxidants in biology; Antioxidants in disease and oxidative stress.

Choice: American College Library Association (Dr N. Duran, Illinois State University)
"The idea of collectively presenting information about antioxidants in technology and biology is certainly useful. The extensive chapter references provide a historical overview of the literature dating back to early in this century. Many figures, tables and schemes explain chemical and biological reactions."

Free Radical Research Society Newsletter (Professor Ursini, Padova University, Italy):
"Drives the reader through mechanisms of peroxidation and antioxidant effect to discuss possible interventions in polymer technology. Chemical aspects are described in depth for the biologist interested to rationalise the effect of enzymes, drugs and food components interacting with hydroperoxides."

Polymer Degradation and Stability (Dr N. Billingham, University of Sussex):
"This distillation of a lifetime of work by one of the world's leading experts. It should find an essential place on the desk of anyone who is interested in the oxidative deterioration of organic materials and the ways of stopping it. A must have for anyone in the field."

Chemistry and Industry (Garry Duthie, Rowett Research Institute, Aberdeen):
"Professor Scott's considerable expertise comes through clearly in chapters on peroxidation in chemistry and chemistry technology, chain-breaking antioxidants, preventive antioxidants, synergism and technical performance. A welcome addition to the fields of free radicals, will hopefully generate healthy debate between the classical oxidation chemists and those in fields of biology, nutrition and medicine."

OBJECT-ORIENTED TECHNOLOGY AND COMPUTING SYSTEMS RE-ENGINEERING

H. ZEDAN and A. CAU, Software Technical Research Laboratory, De Montfort University, Leicester

ISBN: 1-898563-56-X   *ca.* 200 pages   1998

This book delivers, for advanced study and R&E in computing science and electronics engineering, the latest developments in object technology and their impact in computing systems re-engineering; Object-oriented programming is here shown to provide support for constructing large scale systems, cheaply built and with re-usable components, adaptable to changing requirements by efficient and cost-effective techniques.

The contributing authors, internationally recognised authorities from Finland, France, Germany, Italy, Poland, Spain, UK and USA, apply industrial techniques and structured object-oriented methodologies to forward and reverse engineering of computer systems. This book takes stock of progress of that work showing its promise and feasibility, and how its structured technology can overcome the limitations of forward engineering methods used in industry. Forward methods are focused in the domain of reverse engineering to implement a high level of specification for existing software.

Edited by Professor Zedan and Dr Cau, the text here presented is the selected content of the first UK Colloquium on Object Technology and Systems Re-engineering held at Oxford University. The conference was sponsored by British Telecom Laboratories, EMSI Limited, TriReme International, and the OOSP specialised Group of BCS.

Contents: Formal Methods in OO: Towards an object-oriented design methodology for hybrid systems; Fair objects; System Re-engineering: Re-engineering requirements specifications for reuse: A synthesis of 3 years industrial experience; Pre-processing COBOL programs for reverse engineering in a software maintenance tool; Re-engineering procedural software to object-oriented software using design transformations and resource usage matrix; Using OO design to enhance procedural software; Devising coexistence strategies for objects with legacy systems; OO Languages: Systems of systems as communicating structures; Applications of OO: Object-oriented development of x-ray spectrometer software; Object-oriented model for expert systems implementation; Design Patterns and CORBA: Design patterns and their role in formal object-oriented development; Suitability of CORBA as a heterogeneous distributed platform.

Printed and bound by CPI Group (UK) Ltd, Croydon, CR0 4YY

22/10/2024

01777397-0002